METHODS IN MOLECULAR BIOLOGY

Series Editor
John M. Walker
School of Life and Medical Sciences
University of Hertfordshire
Hatfield, Hertfordshire, UK

For further volumes:
http://www.springer.com/series/7651

For over 35 years, biological scientists have come to rely on the research protocols and methodologies in the critically acclaimed *Methods in Molecular Biology* series. The series was the first to introduce the step-by-step protocols approach that has become the standard in all biomedical protocol publishing. Each protocol is provided in readily-reproducible step-by-step fashion, opening with an introductory overview, a list of the materials and reagents needed to complete the experiment, and followed by a detailed procedure that is supported with a helpful notes section offering tips and tricks of the trade as well as troubleshooting advice. These hallmark features were introduced by series editor Dr. John Walker and constitute the key ingredient in each and every volume of the *Methods in Molecular Biology* series. Tested and trusted, comprehensive and reliable, all protocols from the series are indexed in PubMed.

Caveolae

Methods and Protocols

Edited by

Cedric M. Blouin

Institut Curie – Centre de Recherche, PSL Research University, Membrane Mechanics and Dynamics of Intracellular Signaling Laboratory, CNRS UMR3666, INSERM U1143, Paris, France

Humana Press

Editor
Cedric M. Blouin
Institut Curie – Centre de Recherche
PSL Research University
Membrane Mechanics and Dynamics of Intracellular Signaling Laboratory
CNRS UMR3666, INSERM U1143, Paris, France

ISSN 1064-3745 ISSN 1940-6029 (electronic)
Methods in Molecular Biology
ISBN 978-1-0716-0731-2 ISBN 978-1-0716-0732-9 (eBook)
https://doi.org/10.1007/978-1-0716-0732-9

Cover image courtesy of Cedric M. Blouin

This Humana imprint is published by the registered company Springer Science+Business Media, LLC part of Springer Nature.
The registered company address is: 1 New York Plaza, New York, NY 10004, U.S.A.

Preface

Caveolae are bulb-shaped plasma membrane invaginations discovered in the 1950s. Since then, these membrane structures have been involved in many essential cellular functions and pathologies. Recent studies have brought new tools to understand their underlying molecular mechanisms. The aim of this issue is to gather protocols and methods to study either the structure or the functions of caveolae and their components in several normal and pathophysiological situations. We hope this collection of chapters will enable researchers, new in the field or not, to discover new roles or regulations of formed caveolae and proteins composing their coat in various model organisms.

The first two chapters provide elegant methods for the fine visualization of caveolae and caveolar proteins with transmission electron microscopy either using the APEX2 peroxidase (Chapter 1) or the freeze fracture technique and immunogold labeling (Chapter 2).

Chapter 3 describes how to study caveolae in the context of primary cilia with signaling assays coupled to immunofluorescence microscopy. In Chapters 4–7, several methods for quantitative image analysis are presented. They provide tools to analyze localization and dynamics of caveolae at the cell surface.

Protein–protein interactions are essential for caveolae functions. Chapter 8 describes the method for isolation of plasma membrane-derived vesicles to study potential interactions with caveolar proteins. The use of biotin proximity labeling as a tool for identification of new interactors is described in Chapter 9. Chapter 10 provides a biochemical assay to determine cavin interactions by immunoprecipitation and pull-down. Methods for monitoring caveolae protein interactions with membrane lipids are also outlined in Chapters 11 and 12, which describe different protocols for liposome binding assays with co-sedimentation (Chapter 11) or flotation (Chapter 12). Chapter 13 describes how to produce recombinant caveolin protein for further structural determinations by NMR spectroscopy.

The CRISPR/Cas9 technology has revolutionized gene editing and gives researchers valuable tools to deplete or tag endogenous proteins. Chapter 14 provides protocols to use this strategy for caveolae protein studies.

Since caveolae are essential plasma membrane organelles for mechanoprotection, the following chapters introduce several methods to study caveolae functions in cell mechanics. A protocol to monitor membrane tension using optical tweezers is described in Chapter 15, and methods to follow cellular response to mechanical stress by live imaging in vivo (Chapter 16) or on micropatterned substrate (Chapter 17) are also presented.

To illustrate the relevance of caveolae study in pathophysiology, Chapter 18 finally describes the immunohistochemistry and immunofluorescence protocols and analysis for the muscle caveolin protein, caveolin-3, in the diagnostics of neuromuscular diseases.

Paris, France *Cedric M. Blouin*

Contents

Contributors

GEORGE S. BAILLIE • *Institute of Cardiovascular and Medical Sciences, College of Medical, Veterinary and Life Sciences, University of Glasgow, Glasgow, UK*

CEDRIC M. BLOUIN • *Institut Curie – Centre de Recherche, PSL Research University, Membrane Mechanics and Dynamics of Intracellular Signaling Laboratory, CNRS UMR3666, INSERM U1143, Paris, France*

ZHENLONG CHEN • *Department of and Anesthesiology, University of Illinois at Chicago, Chicago, IL, USA*

SØREN T. CHRISTENSEN • *Department of Biology, University of Copenhagen, Copenhagen Ø, Denmark*

RAIBATAK DAS • *Applied BioMath, Concord, MA, USA*

MELISSA DEWULF • *Institut Curie – Centre de Recherche, PSL Research University, Membrane Mechanics and Dynamics of Intracellular Signaling Laboratory, CNRS UMR3666, INSERM U1143, Paris, France*

TOYOSHI FUJIMOTO • *Department of Anatomy and Molecular Cell Biology, Nagoya University Graduate School of Medicine, Nagoya, Japan*

KERNEY JEBRELL GLOVER • *Department of Chemistry, Lehigh University, Bethlehem, PA, USA*

THOMAS E. HALL • *Institute for Molecular Bioscience, The University of Queensland, Brisbane, Australia*

DENISA HATHAZI • *Department of Clinical Neurosciences, University of Cambridge, Cambridge, UK*

TAKASHI HIRAMA • *Department of Thoracic Surgery, Institute of Development, Aging and Cancer, Tohoku University, Sendai, Miyagi, Japan; Program in Cell Biology, The Hospital for Sick Children, Toronto, ON, Canada; Division of Respirology, Department of Medicine, University of Toronto, Toronto, ON, Canada*

MADLEN HUBERT • *Department of Integrative Medical Biology, Umeå University, Umeå, Sweden*

JEFFREY A. JULIEN • *Department of Chemistry, Lehigh University, Bethlehem, PA, USA*

MICHAEL M. KESSELS • *Institute for Biochemistry I, Jena University Hospital – Friedrich Schiller University Jena, Jena, Germany*

DARIUS V. KÖSTER • *Centre for Mechanochemical Cell Biology and Division of Biomedical Sciences, Warwick Medical School, University of Warwick, Coventry, UK*

ELIN LARSSON • *Department of Integrative Medical Biology, Umeå University, Umeå, Sweden*

YE-WHEEN LIM • *Institute for Molecular Bioscience, The University of Queensland, Brisbane, Australia*

HARRIET P. LO • *Institute for Molecular Bioscience, The University of Queensland, Brisbane, Australia*

ALEXANDER LUDWIG • *School of Biological Sciences, NTU Institute of Structural Biology, Nanyang Technological University, Singapore, Singapore*

RICHARD LUNDMARK • *Department of Integrative Medical Biology, Umeå University, Umeå, Sweden*

CAROLINA MENDOZA-TOPAZ • *MRC Laboratory of Molecular Biology, Cambridge, UK*

Richard D. Minshall • *Department of Pharmacology, University of Illinois at Chicago, Chicago, IL, USA; Department of and Anesthesiology, University of Illinois at Chicago, Chicago, IL, USA*

Johanne B. Mogensen • *Department of Biology, University of Copenhagen, Copenhagen Ø, Denmark; Department of Hematology, Biotech Research and Innovation Center, Copenhagen N, Denmark*

Stine K. Morthorst • *Department of Biology, University of Copenhagen, Copenhagen Ø, Denmark; Department of Hematology, Biotech Research and Innovation Center, Copenhagen N, Denmark*

Timothy M. Palmer • *Centre for Atherothrombosis and Metabolic Disease, Hull York Medical School, University of Hull, Hull, UK*

Robert G. Parton • *Institute for Molecular Bioscience, The University of Queensland, Brisbane, Australia; Centre for Microscopy and Microanalysis, The University of Queensland, Brisbane, Australia*

Lotte B. Pedersen • *Department of Biology, University of Copenhagen, Copenhagen Ø, Denmark*

Sarah M. Plucinsky • *Department of Chemistry, Lehigh University, Bethlehem, PA, USA*

Joanna Podkalicka • *Laboratoire Physico-Chimie Curie, Institut Curie, PSL Research University, Sorbonne Université, CNRS UMR168, Paris, France; Institut Curie – Centre de Recherche, PSL Research University, Membrane Mechanics and Dynamics of Intracellular Signaling Laboratory, CNRS UMR3666, INSERM U1143, Paris, France; Laboratory of Cytobiochemistry, Faculty of Biotechnology, University of Wrocław, Wrocław, Poland*

Britta Qualmann • *Institute for Biochemistry I, Jena University Hospital – Friedrich Schiller University Jena, Jena, Germany*

Andreas Roos • *Department of Neuropediatrics, Developmental Neurology and Social Pediatrics, Centre for Neuromuscular Disorders in Children, University Hospital Essen, University of Duisburg-Essen, Essen, Germany*

Ulrike Schara • *Department of Neuropediatrics, Developmental Neurology and Social Pediatrics, Centre for Neuromuscular Disorders in Children, University Hospital Essen, University of Duisburg-Essen, Essen, Germany*

Eric Seemann • *Institute for Biochemistry I, Jena University Hospital – Friedrich Schiller University Jena, Jena, Germany*

Yosuke Senju • *Research Institute for Interdisciplinary Science (RIIS), Okayama University, Okayama, Japan*

Elena Shvets • *MRC Laboratory of Molecular Biology, Cambridge, UK*

Miriam Stoeber • *Department of Cell Physiology and Metabolism, University of Geneva, Geneva, Switzerland*

Shiro Suetsugu • *Division of Biological Science, Nara Institute of Science and Technology, Nara, Japan*

Takashi Takahashi • *Division of Molecular Carcinogenesis, Center for Neurological Diseases and Cancer, Nagoya University Graduate School of Medicine, Nagoya, Japan*

Peter T. Toth • *Department of Pharmacology, University of Illinois at Chicago, Chicago, IL, USA; Research Resources Center Fluorescence Imaging Core, University of Illinois at Chicago, Chicago, IL, USA*

Jamie J. L. Williams • *Institute of Molecular, Cell and Systems Biology, College of Medical, Veterinary and Life Sciences, University of Glasgow, Glasgow, UK*

TOMOYA YAMAGUCHI • *Department of Cancer Biology, Graduate School of Medical Sciences, Kumamoto University, Kumamoto, Japan; Division of Molecular Carcinogenesis, Center for Neurological Diseases and Cancer, Nagoya University Graduate School of Medicine, Nagoya, Japan*
ADRIANA M. ZIMNICKA • *Department of Pharmacology, University of Illinois at Chicago, Chicago, IL, USA*

Chapter 1

Selective Visualization of Caveolae by TEM Using APEX2

Alexander Ludwig

Abstract

Caveolae are small flask- or cup-shaped invaginations of the plasma membrane found in almost all vertebrate cells. Due to their small size (50–100 nm), transmission electron microscopy (TEM) has been the method of choice to study caveolae formation and ultrastructure and, more recently, to resolve the sub-caveolar localization of its protein components using novel protein labeling methods for TEM. This chapter describes a protocol for the selective visualization of caveolae and caveolar proteins by TEM, 3D tomography, and correlative light and electron microscopy (CLEM) using the peroxidase APEX2.

Key words Caveolae, APEX2, TEM, 3D tomography, Correlative light and electron microscopy, Caveolin, Cavin

1 Introduction

Caveolae are a ubiquitous feature of almost all vertebrate cells and so have fascinated cell biologists since the early days of EM [1–3]. Caveolae are particularly abundant in endothelial cells, muscle cells, and adipocytes, in which they can cover up to 50% of the cell surface. They are also prominent in epithelial cells and fibroblasts but are absent from neurons and most blood cells. Apart from forming characteristic flask- or cup-shaped invaginations in the plasma membrane, caveolae also exist as more complex interconnected networks and clusters, often referred to as caveolar rosettes.

Caveolae are composed of two major protein components; the caveolin family (caveolin-1, -2, and -3) and the cavin family (cavin1, 2, 3, and 4) [4, 5]. Caveolin-1 and cavin1 are essential for generating caveolae in vivo, and they are the key structural determinants of the caveolar membrane coat [6–9]. The organization of the caveolar coat has been studied using different EM techniques [10–13] and EM labeling methods [6, 7, 14, 15]. The combined data indicate that the caveolar bulb is formed by membrane-embedded oligomers of caveolins that are stabilized by a peripheral, spiral-like, or striated coat composed of oligomers of cavins. There is also

Cedric M. Blouin (ed.), *Caveolae: Methods and Protocols*, Methods in Molecular Biology, vol. 2169,
https://doi.org/10.1007/978-1-0716-0732-9_1, © Springer Science+Business Media, LLC, part of Springer Nature 2020

evidence that the caveolar coat adopts a polyhedral-like shape [7, 8, 16], yet detailed structural information of how caveolins and cavins generate caveolae is only beginning to emerge.

This chapter describes the use of APEX2 as a tool for the direct visualization of caveolin-1 and cavin1 in cultured cell monolayers by TEM. APEX2 is a modified ascorbate peroxidase that serves as a genetically encoded reporter for TEM [17–19]. APEX2 is 28 kDa in size and can be fused to the N- or C-terminus of proteins. A brief incubation of fixed APEX2 expressing cells in buffer containing hydrogen peroxide (H_2O_2) and 3,3′-diaminonbenzidine (DAB) results in the local oxidation of DAB into an insoluble osmiophilic polymer. This polymer is then made electron-dense during the postfixation step with osmium tetroxide, resulting in a discrete stain that is visible in the TEM.

One advantage of APEX2 over conventional localization techniques (such as immuno-gold EM) is that labeling is performed in situ, without the need for detergents or prior sectioning. In addition, the production of the DAB-osmium deposit is spatially confined and essentially free of background, allowing proteins to be localized effectively and specifically with a precision of approximately 10 nm. APEX2, therefore, also overcomes some inherent technical problems often encountered in immuno-gold EM, including poor or insufficient labeling due to epitope masking and nonspecific labeling due to antibody cross-reactivity. This makes APEX2 an ideal tool for protein visualization by 2D TEM, 3D tomography, CLEM, and EM volume imaging (such as serial block-face SEM or focused ion beam SEM). Finally, unlike horseradish peroxidase (HRP), APEX2 is active both in the lumen/extracellular space as well as in the cytoplasm, which permits labeling of essentially any subcellular compartment of interest. It is also worth pointing out that APEX2 has become a popular tool for proximity biotinylation [20]. Here, APEX2 catalyzes the oxidation of biotin phenol into a highly reactive phenoxyl radical, which reacts with electron-dense amino acid side chains (mostly tyrosines) in an estimated radius of 20 nm. Biotinylated proteins can then be isolated and analyzed by mass spectrometry. Hence, APEX2 is a versatile reporter for TEM and proximity proteomics.

To study caveolae by CLEM, we have generated caveolin-1 and cavin-1 fusion proteins with a C-terminal tandem APEX2-EGFP tag. Both fusion proteins localize to caveolae when transfected into cultured cells [7, 15]. Expression of the caveolin-1-APEX2-EGFP protein in *caveolin-1 −/−* mouse embryonic fibroblasts (MEFs) rescues caveolae formation, indicating that the fusion protein is functional [7]. Here we describe a standard, room-temperature EM protocol for cell monolayers transiently or stably transfected

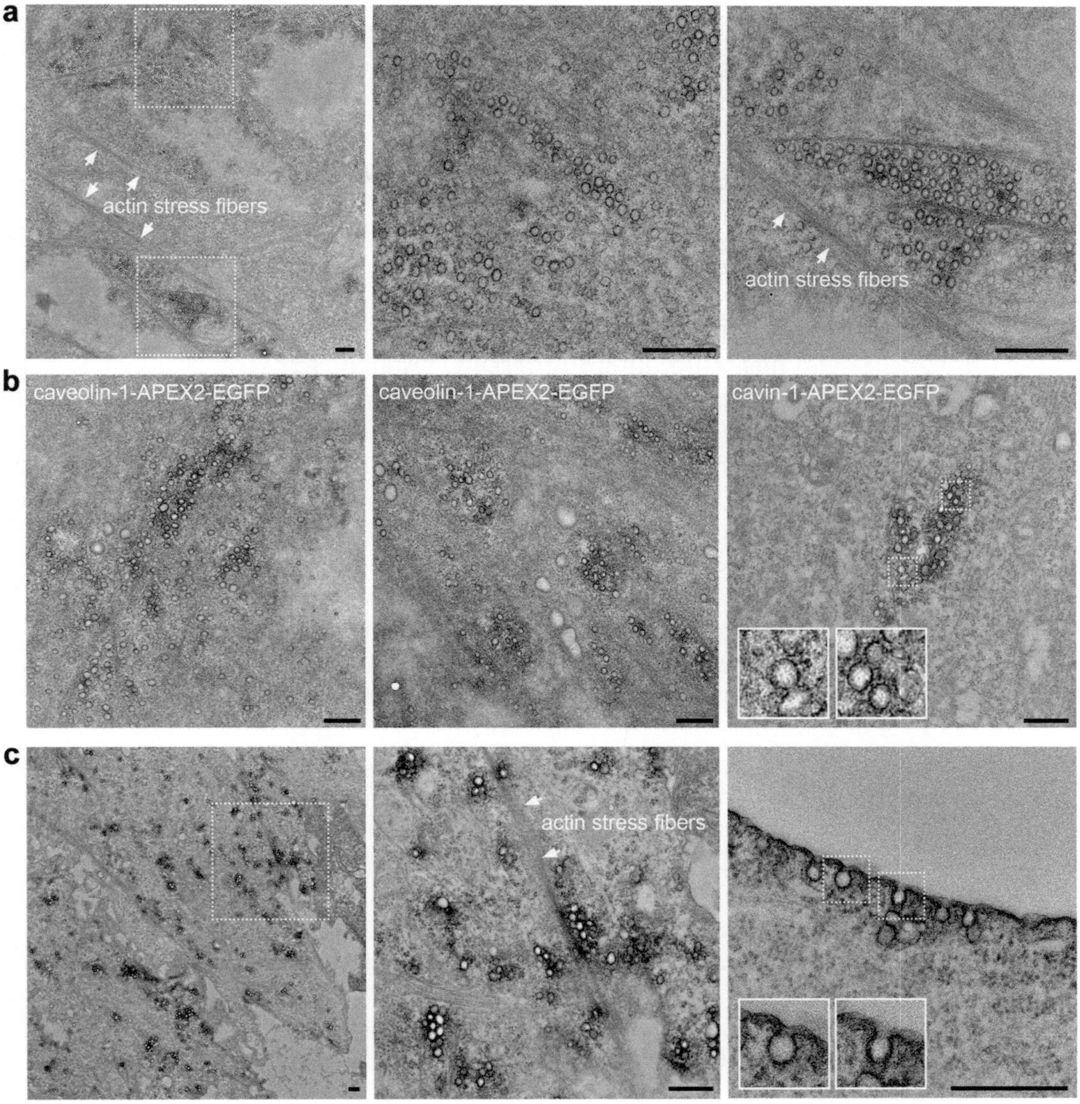

Fig. 1 Labeling of caveolae for transmission electron microscopy using APEX2. (**a**) Caveolin-1-APEX2-EGFP transiently transfected into caveolin-1 −/− iMEFs. (**b**) Caveolin-1 or cavin-1 APEX2-EGFP fusion proteins transiently transfected into HeLa cells. (**c**) Cavin-1-APEX2-EGFP stably transfected into RPE1 cells. All images are from 70–80 nm cell sections recorded at 120 kV. Insets are magnified views of the boxed areas. All scale bars are 500 nm

with cavin-1-APEX2-EGFP or caveolin-1-APEX2-EGFP (Fig. 1) (*see* **Note 1**). The protocol is applicable to other APEX2 fusion proteins and is suitable for 3D tomography and CLEM experiments.

2 Materials

Essentially all chemicals used in the following protocol are harmful to humans and the environment. Particular care should be taken in handling cacodylate buffer (toxic, arsenic), uranyl acetate (radioactive), DAB (carcinogenic), and osmium tetroxide (highly volatile, toxic, and strong oxidizer). Gloves should be worn at all times, and all buffer exchanges and incubation steps should be performed in a fume hood. Chemicals should be collected and disposed of according to standard operating procedures. Consult the MSDS documents if you are uncertain. Osmium tetroxide waste should be neutralized by addition of 2 volumes of corn oil. Uranyl acetate waste should be collected in a dedicated radioactive waste container. Resin waste and all materials that have been in touch with resin should be collected separately and polymerized at 60 °C for 48 h before disposal. Chemicals and buffers should be purchased in EM grade from vendors such as Electron Microscopy Sciences (EMS). Most buffers and solutions should be prepared immediately before use in double distilled and autoclaved water (ddH_2O). The volumes indicated in the following protocol are sufficient for the simultaneous processing of two samples.

2.1 Cell Culture and Transfection

HeLa cells and MEFs are maintained in Dulbecco's Modified Eagle Medium (DMEM), 10% fetal calf serum (FCS), and Penicillin/Streptomycin at 37 °C, 5% CO_2. RPE1 cells are grown in DMEM/F12 medium supplemented with FCS and antibiotics. For TEM of cell monolayers, plate cells onto MatTek glass-bottom dishes (No 1.5 cover glass, MatTek Corporation P35G-1.5-14-C). It is recommended to coat the glass with fibronectin or poly-L-lysine. For fibronectin-coating, add 150 μl of a 5 μg/ml fibronectin solution prepared in PBS into the center of the MatTek dish, distribute evenly, and store in the incubator for at least 1 h. Aspirate the fibronectin solution, rinse dish briefly with 2 ml of PBS and immediately add the cell suspension. Unless stable cell lines are being used, transfect cells the day after using conventional protocols (*see* **Note 2**). Fix cells for APEX2-TEM 16–24 h post-transfection. Make sure cells are no more than 80% confluent on the day of the experiment.

2.2 Electron Microscopy

1. **Cacodylate buffer (CB) stock solution**

   0.3 M CB: Weigh in 64.2 g sodium cacodylate trihydrate. Dissolve by addition of 800 ml ddH_2O. Adjust to pH 7.4 with 0.2 N HCl and bring to a final volume of 1 l. This is a 3× stock solution. Do not autoclave. Store at 4 °C.

2. **Blocking solution**

   50 mM glycine in 0.1 M CB. To prepare 25 ml blocking solution, add 10 mg of glycine to 0.1 M CB. Let dissolve and store on ice.

3. **Fixative**

   2.5% glutaraldehyde (GA) in 0.1 M CB: Add 1 ml of 25% EM grade GA to 9 ml of 0.1 M CB. Single-use aliquots of GA can be stored at −80 °C. Supplement fixative with 2 mM $CaCl_2$. Prepare fixative fresh and store at room temperature until use.

4. **DAB solution**

   Dissolve 5.4 mg of the 3,3′-Diaminobenzidine (DAB) free base (Sigma D8001) in 1 ml of 0.1 M HCl. Note that alternative sources of DAB are not recommended. Once dissolved, dilute the DAB stock 1:10 in chilled 0.1 M CB buffer (final DAB concentration of 0.54 mg/ml). Store on ice. Add H_2O_2 at a final concentration of 0.03% (w/v) (8.8 mM), and immediately filter onto cells (using a 0.22 μm filter). Alternatively, use a final H_2O_2 concentration of 0.5 mM. In our hands, this results in higher sensitivity and stronger EM contrast (*see* **Note 3**).

5. **Osmium tetroxide (OsO_4) solution**

   1% OsO_4 in 0.1 M CB. In the fume hood and with great care, open an ampoule containing 2 ml of 2% OsO_4 aqueous solution. Aspirate OsO_4 solution with a fine plastic transfer pipette and add to 2 ml of 0.2 M CB. Mix well by pipetting up and down. Supplement with 2 mM $CaCl_2$. Prepare fresh and store on ice in the dark until use.

   Optional Postfixation and EM Contrasting Steps (*See* **Note 4**)

6. **Potassium ferricyanide ($K_4(Fe(CN)_6)$)** enhances general membrane contrast when added *together* with OsO_4. Add 1–2% (w/v) $K_4(Fe(CN)_6)$ to 1% OsO_4 in 0.1 M CB.

7. **Tannic acid (TA)** enhances general membrane contrast when applied *after* the OsO_4 postfixation step. Use low molecular weight TA $(C_{14}H_{10}O_9)_n$. Prepare a 1–2% TA (w/v) solution in 0.1 M CB immediately before use. Higher TA concentration and longer incubation times result in stronger membrane contrast.

8. **Uranyl acetate (UA)** enhances EM contrast of proteins and nucleic acids. Prepare a 1% UA solution in ddH_2O from a 4% (w/v) UA stock (made from uranyl acetate dihydrate). Filter (0.22 μm) just prior to use. Incubate on ice in the dark for 1 h. Alternatively, stain overnight at 4 °C.

9. Durcupan™ ACM resin
   Durcupan™ ACM resin is a four-component epoxy resin. Using a clean plastic beaker weigh in components A-D in the following order. Be precise:

   Component A: 11.4 g (decant).

   Component B: 10 g (decant).

   Component C: 0.3 g (use a 2 ml plastic transfer pipette).

   Component D: 0.1 g (use a 1 ml plastic transfer pipette).

   Stir slowly but thoroughly for 5 min using a clean plastic stick. Avoid air bubbles. Seal with parafilm and store in the dark for 30 min. Prepare 1:2 and 2:1 resin:anhydrous ethanol solutions. Prepare 6 ml of each. Mix thoroughly.

3 Methods

3.1 Fixation, Light Microscopy and APEX2 Reaction

1. To fix cells grown in MatTek glass-bottom dishes aspirate or decant the medium and immediately add 2 ml of room temperature fixative. Decant, add fresh fixative and leave at room temperature for 5 min.
2. Move dish onto ice, and keep on ice in the dark for 1 h—all subsequent steps are carried out on ice until infiltration in resin.
3. Rinse cells 3 × 5 min in chilled 0.1 M CB.
4. Block unreacted aldehydes with blocking solution: 3 × 10 min (this step can be omitted if light microscopy is not part of the workflow).
5. Rinse 3 × 2 min in chilled 0.1 M CB.
6. For CLEM, place the MatTeK dish onto a microscope stage cooled to 4 °C (if available). Mark the orientation of the dish using a permanent pen. Capture fluorescence images of regions of interest (ROI) using standard GFP filter sets (confocal z-stacks are preferable). Use a high NA 40× oil or 63× oil lens, if available. Capture transmitted light images (pre-DAB images). Safe ROIs in the software (if a stage position tool is available).
7. Gently aspirate CB and immediately filter the DAB solution onto the cells. Make sure the location of the dish on the stage is not altered.
8. Incubate cells for 0.5–5 min in DAB solution (*see* **Note 5**).
9. To halt the reaction, aspirate DAB solution and rinse cells 5 × 2 min with chilled 0.1 M CB.
10. Capture transmitted light images of ROIs (post-DAB images). A brown reaction product (insoluble DAB polymer) should be visible.

3.2 Postfixation, Counterstaining, Dehydration, and Embedding

1. Postfix cells in 1% OsO_4 solution for 1 h on ice in the dark.
2. Optional: The following steps can be included in the protocol—alone or in combination—to increase EM contrast of membranes, nucleic acids, and proteins (*see* **Note 4**).
   (a) To increase membrane contrast, $K_4(Fe(CN)_6)$ can be added to 1% OsO_4 solution at a final concentration of 0.5–2% (w/v). Incubate on ice and in the dark for 1 h.
   (b) After osmification in 1% OsO_4 solution, membrane contrast can be enhanced by incubation with 1–2% tannic acid solution. Incubate 30–60 min at room temperature in the dark.
   (c) General EM contrast can be enhanced by staining cells with 1% UA solution for 1 h on ice or overnight in the fridge in the dark.
3. Rinse cells 5 × 2 min in chilled ddH_2O.
4. Dehydration: Dehydrate cells using a chilled graded ethanol series (20%, 50%, 70%, 90%) for 3 min each. Make sure you work quickly to prevent cells from drying out.
5. Remove sample from ice and rinse 3 × 3 min in room temperature anhydrous (100%) ethanol. Wipe off any water/liquid from the outside of the dish and the lid to prevent residual water in your sample.
6. Resin infiltration: To facilitate infiltration the sample is first infiltrated with an ascending series of resin/ethanol solution:
   (a) 1:2 (v/v) resin:100% ethanol for 30–60 min
   (b) 2:1 (v/v) resin:100% ethanol for 30–60 min
   (c) 100% resin overnight. Remove the lid from the MatTek dish and cover the dish to prevent dust settling onto the cells and in the resin
   (d) The following day, prepare fresh resin and perform 3–6 additional resin exchanges. Infiltration and resin exchange can be greatly facilitated by incubating the sample in an oven set to no more than 45 °C. Exchange resin every 45 min.
7. Polymerize resin at 60 °C for 48 h.

3.3 Relocation of ROIs, Mounting, and Ultramicrotomy

1. Relocate the areas of interest using a binocular or a tissue culture light microscope. Mark the locations on the backside of the resin (not on the glass) with a permanent pen. Fix the dish in a small bench vice and, using a jewelers saw, carefully cut out the area of interest. Free the block from resin debris with water, remove pen mark from the back of the block with a piece of tissue soaked in 100% ethanol, and mount sample on a resin pin (dummy) with superglue. Make sure cells are facing up.

2. Typically, the glass coverslip remains attached to the block. To dislodge it, insert the block into the ultramicrotome sample holder and add a small drop of water onto the glass surface. Place a sharp razor blade at the edge of the block face, just below the glass, at a slight angle. Apply some pressure and carefully snap the coverslip off the block face. Clean the block face with a compressed air duster. Make sure no glass splinters are left on the block face.
3. Using a razor blade and a glass or trimming knife, trim the block face to a trapezium shape of approximately 1 mm × 1 mm.
4. Section sample using a diamond knife. Align the knife edge exactly parallel to the block face and collect the very first and all subsequent sections (serial sections). For 2D TEM cut 60–80 nm sections, for tomography cut 200–300 nm sections.
5. Collect sections on EM grids of your choice. Glow discharge grids just prior to using a plasma cleaner (e.g. Cressington 208, set to 5 V for 30 s). For CLEM, it is recommended to use formvar and carbon coated 1 mm slot grids. For tomography, use 100 mesh copper grids.
6. Sections picked up on plain mesh grids should be coated on both sides with a 2–3 nm thick carbon film using a carbon coater.
7. For tomography, dip the grids for 30–60 s into a solution of 5 or 10 nm gold particles diluted in PBS, 0.02% (w/v) BSA. Remove excess liquid with a small piece of blotting paper and let the grids air-dry. Aim for approximately 20–30 gold particles in a field of view.

4 Notes

1. In aldehyde-fixed and conventionally processed TEM samples caveolae typically appear as flask-shaped membrane invaginations (with a narrow neck). However, it has been shown that their shape is more cup-like (with a wider neck) in cells fixed by high-pressure freezing and subsequently processed by freeze-substitution (HPF-FS) [13]. This suggests that particular structural features of caveolae are sensitive to chemical fixation or dehydration. The method described here uses a conventional EM protocol, in which aldehyde-fixed cells are dehydrated, stained, and embedded at room temperature or on ice. A more labor- and time-intensive protocol that combines APEX2 labeling with HPF-FS has recently been described [21].

2. It is highly recommended to generate cell lines stably expressing APEX2 fusion proteins of caveolin-1 or cavin1. Preferably, the fusion proteins should be expressed at or below endogenous levels. This can be achieved through conventional transfection protocols and clonal selection or through genome editing.
3. It is important to consider that APEX [18] and APEX2 [17] are inhibited by H_2O_2. It has been proposed that APEX2 is superior to APEX for EM imaging and proximity biotinylation because it is less susceptible to H_2O_2-induced inhibition [17]. Our experience is that although APEX2 is clearly more sensitive than APEX, even APEX2 fails to produce sufficient EM contrast in cells expressing relatively low levels of fusion proteins. Interestingly, we found that lowering the H_2O_2 concentration to 0.5 mM results in a marked increase in EM contrast, and hence improves the detection of APEX2 fusion proteins expressed at lower levels. We speculate that lower H_2O_2 concentrations allow the enzyme to be active for longer, enhancing the sensitivity of the method. Although we have not systematically tested this as yet, it seems plausible then that the amount of EM contrast may be adjustable (to a certain extent) by varying the H_2O_2 concentration within the 10–0.5 mM range.
4. It is recommended to perform initial APEX2 experiments without any EM counterstains. This facilitates interpretation of the stain and greatly helps in assessing whether the observed EM contrast is caused specifically by APEX2. Once specificity of the EM label has been ascertained, additional stains can be included in the protocol to increase contrast of proteins, nucleic acids, and/or lipids. In our hands a short (30–60 min) incubation with 1–2% (w/v) tannic acid (subsequent to postfixation with osmium tetroxide) noticeably enhances membrane contrast without masking the APEX2-mediated DAB-osmium deposit (*see* [22] for a detailed description of how tannic acid imparts membrane contrast). A similar effect is achieved by the addition of $K_4(Fe(CN)_6)$ during postfixation with osmium tetroxide. En block staining with 1% uranyl acetate (alone or in combination with tannic acid) is also compatible with the protocol.
5. APEX2-mediated oxidation of DAB is an extremely rapid process that occurs within tens of seconds. However, due to the inhibitory effect of H_2O_2 (*see* **Note 3**) the reaction is essentially completed within minutes (or sooner). Hence, the intensity of the EM stain is mostly determined by the amount of APEX2 protein expressed and cannot easily be controlled by adjusting the labeling time. It is advisable, therefore, to empirically determine optimal protein expression levels and labeling times.

Excessive overexpression and long labeling times should be avoided as both of which can result in a more diffuse stain and partial degradation of ultrastructure.

Acknowledgments

This work was supported by grants from MOE Singapore (TIER1 RG39-14) and Nanyang Technological University (NIM/03/2016).

References

1. Stan RV (2005) Structure of caveolae. Biochim Biophys Acta 1746:334–348
2. Parton RG, Del Pozo MA (2013) Caveolae as plasma membrane sensors, protectors and organizers. Nat Rev Mol Cell Biol 14:98–112
3. Shvets E, Ludwig A, Nichols BJ (2014) News from the caves: update on the structure and function of caveolae. Curr Opin Cell Biol 29:99–106
4. Kovtun O, Tillu VA, Ariotti N et al (2015) Cavin family proteins and the assembly of caveolae. J Cell Sci 128:1269–1278
5. Hansen CG, Nichols BJ (2010) Exploring the caves: cavins, caveolins and caveolae. Trends Cell Biol 20:177–186
6. Ludwig A, Howard G, Mendoza-Topaz C et al (2013) Molecular composition and ultrastructure of the caveolar coat complex. PLoS Biol 11:e1001640
7. Ludwig A, Nichols BJ, Sandin S (2016) Architecture of the caveolar coat complex. J Cell Sci 129:3077–3083
8. Stoeber M, Schellenberger P, Siebert CA et al (2016) Model for the architecture of caveolae based on a flexible, net-like assembly of Cavin1 and Caveolin discs. Proc Natl Acad Sci U S A 113:E8069–E8078
9. Kovtun O, Tillu VA, Jung W et al (2014) Structural insights into the organization of the cavin membrane coat complex. Dev Cell 31:405–419
10. Peters KR, Carley WW, Palade GE (1985) Endothelial plasmalemmal vesicles have a characteristic striped bipolar surface structure. J Cell Biol 101:2233–2238
11. Lebbink MN, Jimenez N, Vocking K et al (2010) Spiral coating of the endothelial caveolar membranes as revealed by electron tomography and template matching. Traffic 11:138–150
12. Rothberg KG, Heuser JE, Donzell WC et al (1992) Caveolin, a protein component of caveolae membrane coats. Cell 68:673–682
13. Richter T, Floetenmeyer M, Ferguson C et al (2008) High-resolution 3D quantitative analysis of caveolar ultrastructure and caveola-cytoskeleton interactions. Traffic 9:893–909
14. Gambin Y, Ariotti N, Mcmahon KA et al (2014) Single-molecule analysis reveals self assembly and nanoscale segregation of two distinct cavin subcomplexes on caveolae. eLife 3: e01434
15. Ludwig A, Nguyen TH, Leong D et al (2017) Caveolae provide a specialized membrane environment for respiratory syncytial virus assembly. J Cell Sci 130(6):1037–1050
16. Walser PJ, Ariotti N, Howes M et al (2012) Constitutive formation of caveolae in a bacterium. Cell 150:752–763
17. Lam SS, Martell JD, Kamer KJ et al (2015) Directed evolution of APEX2 for electron microscopy and proximity labeling. Nat Methods 12:51–54
18. Martell JD, Deerinck TJ, Sancak Y et al (2012) Engineered ascorbate peroxidase as a genetically encoded reporter for electron microscopy. Nat Biotechnol 30:1143–1148
19. Martell JD, Deerinck TJ, Lam SS et al (2017) Electron microscopy using the genetically encoded APEX2 tag in cultured mammalian cells. Nat Protoc 12:1792–1816
20. Hung V, Udeshi ND, Lam SS et al (2016) Spatially resolved proteomic mapping in living cells with the engineered peroxidase APEX2. Nat Protoc 11:456–475
21. Tsang TK, Bushong EA, Boassa D et al (2018) High-quality ultrastructural preservation using cryofixation for 3D electron microscopy of genetically labeled tissues. eLife 7
22. Wagner RC (1976) The effect of tannic acid on electron images of capillary endothelial cell membranes. J Ultrastruct Res 57:132–139

Chapter 2

Freeze-Fracture Replica Immunolabeling of Cryopreserved Membrane Compartments, Cultured Cells and Tissues

Eric Seemann, Michael M. Kessels, and Britta Qualmann

Abstract

Membrane topology information and views of membrane-embedded protein complexes promote our understanding of membrane organization and cell biological function involving membrane compartments. Freeze-fracturing of biological membranes offers both stunning views onto integral membrane proteins and perpendicular views over wide areas of the membrane at electron microscopical resolution. This information is directly assessable for 3D analyses and quantitative analyses of the distribution of components within the membrane if it were possible to specifically detect the components of interest in the membranes. Freeze-fracture replica immunolabeling (FRIL) achieves just that. In addition, FRIL preserves antigens in their genuine cellular context free of artifacts of chemical fixation, as FRIL uses chemically unfixed cellular samples that are rapidly cryofixed. In principle, the method is not limited to integral proteins spanning the membrane. Theoretically, all membrane components should be addressable as long as they are antigenic, embedded into at least one membrane leaflet, and accessible for immunolabeling from either the intracellular or the extracellular side. Consistently, integral proteins spanning both leaflets and only partially inserted membrane proteins have been successfully identified and studied for their molecular organization and distribution in the membrane and/or in relationship to specialized membrane domains. Here we describe the freeze-fracturing of both cultured cells and tissues and the sample preparations that allowed for a successful immunogold-labeling of caveolin1 and caveolin3 or even for double-immunolabelings of caveolins with members of the syndapin family of membrane-associating and -shaping BAR domain proteins as well as with cavin 1. For this purpose samples are cryopreserved, fractured, and replicated. We also describe how the obtained stabilized membrane fractures are then cleaned to remove all loosely attached material and immunogold labeled to finally be viewed by transmission electron microscopy.

Key words Freeze-fracture, Membrane topology, Immunogold labeling, Membrane proteins, Membrane-associated proteins, Nanodomains, Caveolae, Caveolar invaginations, Caveolin, Syndapin

1 Introduction

Freeze-fractured membranes provide undisturbed views onto cellular membrane organizations at magnifications and resolutions high enough to even visualize molecular details when analyzed by electron microscopy [1]. It was this powerful technique in membrane research that strongly supported the fluid-mosaic model in

Cedric M. Blouin (ed.), *Caveolae: Methods and Protocols*, Methods in Molecular Biology, vol. 2169,
https://doi.org/10.1007/978-1-0716-0732-9_2,

membrane biology that prevails until today, as freeze-fracturing was able to directly visualize the organization of components within membranes by transmission electron microscopy (TEM) [2, 3]. The method hereby builds on the tendency of frozen membranes to fracture along the plane through the central hydrophobic core. This splits membranes into half-membrane leaflets [2]. The resulting *en face* views of membranes give spectacular three-dimensional perspectives of cellular organization and details of membrane structure. The success of the freeze-fracturing technique hereby arose from its ability to provide compelling images of membranes and membrane components in an intact, fully preserved cellular context, as chemical fixation artifacts are avoided by rapid cryopreservation [4].

When not general membrane topologies but individual protein components within membranes need to be addressed, however, classical freeze-fracturing methods strictly depend on the ability to instantly recognize a given protein or protein complex of interest based on already existing structural knowledge. In the case of integral membrane proteins, it is even uncertain in which of the two membrane leaflets these proteins will end up after fracturing. As many membrane proteins seem not to display any recognizable characteristic shape at the resolution of transmission electron microscopy (TEM), the clear recognition of membrane proteins therefore represents an important aspect in studying endogenous proteins in their genuine membrane context.

The combination of freeze-fracturing methods and immunolabeling techniques is able to overcome this limitation and to therewith unleash the full scientific potential of freeze-fractured biological specimen. Early work was restricted to labeling of the extracellular surface of cells either prior or subsequent to replication [5–7] and coined the term freeze-fracture immunolabeling (FRIL) for such techniques. A landmark paper by Fujimoto (1995) [8] presented the first immunolabeling of a freeze-fractured tissue sample at the cytosolic side. Fujimoto freeze-fractured chick liver, replicated the two membrane halves by platinum/carbon (Pt/C) evaporation and then probed the Pt/C-free, cytosolic membrane surface with antibodies. Once Fujimoto had cleaned the freeze-fractured Pt/C replica from loosely attached cellular remnants using harsh sodium dodecyl sulfate (SDS) treatments, it was indeed possible to specifically detect the protein occludin at the tight junctions of hepatocytes [8]. The method is based on the fact that molecules in direct contact with the replica film such as membrane lipids and proteins inserted deeply into at least one membrane layer remain associated with the replica cast and the membrane upon the replica cleaning process, whereas all loosely associated or not at all membrane-associated material is removed effectively.

Several recent studies demonstrated that, besides different synaptic receptors [9, 10] and neuronal ion channels [11, 12], also caveolin1 [13–16], caveolin2 [13, 17] and caveolin3 [16], cavin 1 [15, 16] as well as members of the syndapin (PACSIN) family can be detected at the plasma membrane of different cells in culture and even in tissues using FRIL [15, 16, 18]. Coimmunolabelings of caveolin1 with syndapin II, of caveolin1 with cavin1/PTRF, and of cavin1 with syndapin II [15] as well as of caveolin3 with syndapin III at freeze-fracture replica of plasma membrane [16] have also been reported. For all three syndapins, quantitative analyses were used to firmly demonstrate the specific immunolabeling of the endogenous proteins in their genuine membrane environment. Importantly, all of these studies included quantitative evaluations of secondary antibody controls and/or of antibody quench controls with recombinant antigen. They even furthermore included quantitative evaluations of syndapin-deficient samples: the finding that syndapin II can be specifically immunolabeled at membranes of NIH3T3 cells was proven by a comparison of the labeling densities of WT cells in comparison with syndapin II siRNA-treated cells [15]. The specific labeling of syndapin III at membranes of primary cardiomyocytes was demonstrated by quantitative analyses of the anti-syndapin III labeling densities of cardiomyocytes isolated from WT mice in comparison with the labeling densities of cardiomyocytes isolated from syndapin III KO mice [16]. The specific labeling of syndapin I at freeze-fracture replica was even proven by quantitative evaluations of brain tissue material from WT versus syndapin I KO mice [18].

Syndapins [19–21], which exist in one isoform in, e.g., insects and worms, in three isoforms in mammals and in six variants in fish [22], form a subfamily of proteins (for reviews *see* [22–24]) within the superfamily of BAR domain proteins [25–27]. BAR domain proteins are thought to shape membranes by imposing the curvature of the dimeric BAR domain module onto membranes [28]. Syndapins hereby seem to use a molecular mechanism that combines F-BAR domain-mediated membrane scaffolding and insertion of hydrophobic loops located at the concave membrane-binding surface into one membrane leaflet [29].

As also caveolins are thought to insert into one membrane leaflet by hydrophobic and electrostatic interactions [30, 31], it seems very likely that it is the latter property of syndapins, which ensures that syndapin I, syndapin II, and syndapin III all resist the sample preparation procedure of FRIL and are detectable at freeze-fracture replica.

Whether FRIL can likewise be applied to all proteins of the BAR superfamily awaits further work, but we at least thus far demonstrated that FRIL can also be applied to two further F-BAR domain proteins, Cip4/Toca and Nostrin, in in vitro reconstitutions with liposomes [32].

Technically, there are three major steps in generating immunolabeled freeze-fracture replica to be evaluated by transmission electron microscopy (TEM):

- Rapid freezing of the specimen (*see* Subheading 3.1 and 3.2),
- fracturing at low temperature (−140 °C) and generation of replica of the newly exposed frozen surface by carbon and platinum evaporation (in a vacuum, at a certain angle for additional shadowing) (*see* Subheading 3.3) and,
- cleaning of the replica to remove cytosolic material and immunolabeling of membrane-integrated components (*see* Subheading 3.4).

We will in the following go through all of these steps in detail.

For further scientific data and further pictures of FRIL examinations of caveolae, *see* Schlörmann et al. (2010) [33], Koch et al. (2012) [15], and Seemann et al. (2017) [16].

2 Materials

Prepare all solutions using *aqua bidest* and analytical grade reagents. Prepare and store all reagents at room temperature (unless indicated otherwise). Diligently follow all safety and waste disposal regulations.

2.1 Buffers

1. PBS buffer: Phosphate-buffered saline (for 1000 ml)

   Weigh 4.383 g NaCl, 2.208 g $NaH_2PO_4xH_2O$ and 9.255 g $Na_2HPO_4x2H_2O$ into a 1 l glass flask. Add *aqua bidest* to the final volume of 1000 ml. This will yield a solution of 75 mM NaCl, 16 mM $NaH_2PO_4xH_2O$ and 52 mM $Na_2HPO_4x2H_2O$ at pH 7.2.

2. SDS buffer: Sodium dodecyl sulfate (for 100 ml)

   Weigh 2.5 g sodium dodecyl sulfate (SDS), 1.0 g sucrose, and 0.12 g Tris/HCl into a 150 ml glass bulb. Add 90 ml *aqua bidest*, titrate with 1 N HCl to pH 8.4 and add *aqua bidest* to the final volume of 100 ml. Aliquot and store the SDS buffer containing 2.5% (w/v) SDS, 30 mM sucrose, and 10 mM Tris/HCl at pH 8.4 at −20 °C (*see* **Note 1**).

3. LBB buffer: Labeling blocking buffer (for 100 ml)

   Weigh 1.0 g BSA (1% w/v final) and 0.5 g teleostean gelatin (0.5% w/v final) into a glass beaker and add about 50 ml PBS to dissolve the BSA and the gelatin (*see* **Note 2**). Add 5 μl Tween-20 (0.005% final) and then PBS to the final volume. Aliquot the LBB-buffer in 50 ml plastic tubes and store at −20 °C.

4. Fixative: (for 10 ml)

   Add 9.8 ml PBS into a 10 ml plastic tube with lid. Put 200 μl of a 25% v/v glutaraldehyde stock solution (Sigma) into the tube using a syringe (*see* **Note 3**). Close lid and mix carefully to obtain the desired 0.5% v/v glutaraldehyde solution in PBS (1:50 dilution from 25% stock). Store fixative at 4 °C.

2.2 Equipment for Specimen Preparation

1. Freeze-fracture system (BAF 400 T; BAL-TEC, Liechtenstein).
2. Copper sandwich holder (0.6 mm or 0.8 mm profile height) (Fig. 1a).
3. Equipment for cryo-preservation of the samples including,
   (a) a gas bottle with ethane/propane mixture (50:50) with an ascension pipe and a gas pressure regulator,
   (b) a liquid nitrogen-cooled dewar with inner receptable for the propane/ethane mixture (Fig. 1b, c),
   (c) a cooling coil,
   (d) and tubes to connect gas the bottle with the cooling coil.
4. Equipment for interim storage of frozen sample sandwiches in liquid nitrogen (Fig. 1d, e).

2.3 Additional Items

1. Ceramic well plate (ceramic spotting dish).
2. Shaker.
3. Cell scraper.
4. Scalpel.
5. Mesh grids (300–400 mesh).
6. Dewars and liquid nitrogen.
7. Primary antibodies and 5–20 nm (most suitable are 10 nm) gold-labeled secondary antibodies.

3 Methods

Carry out all procedures at room temperature unless stated otherwise.

3.1 Preparation for Cryopreservation of Specimen

1. Put liquid nitrogen as primary coolant into a 1-l dewar (dewar 1) that has space for an inner receptable for 2 ml of ethane/propane mixture.
2. When the liquid nitrogen has cooled down the inner compartment of the dewar, put the inner receptable (Fig. 1b) into the dewar and equilibrate until also the receptable is cooled down.

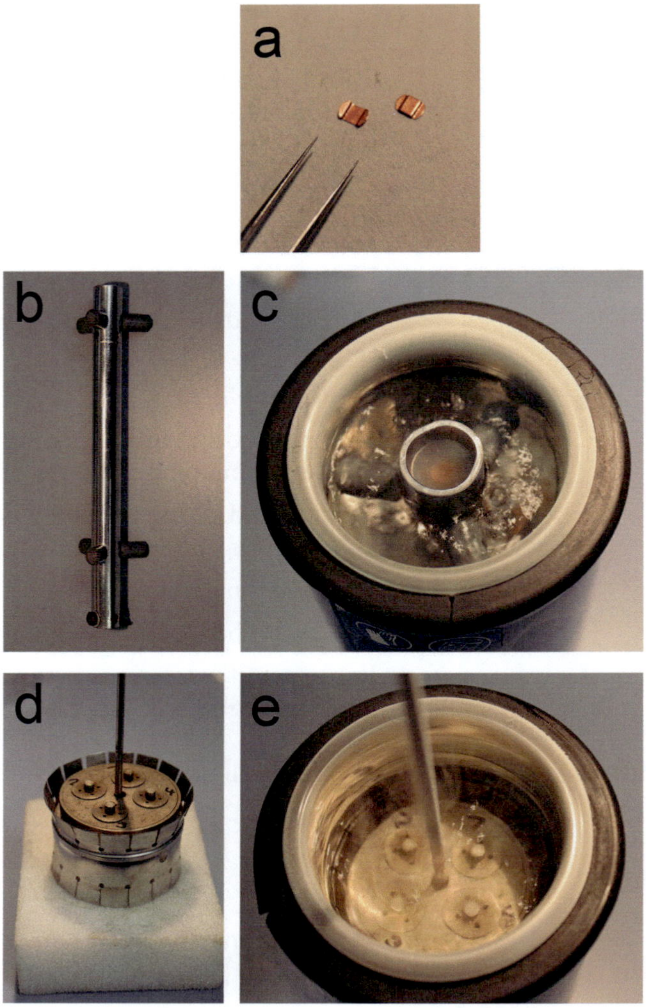

Fig. 1 Equipment needed for cryopreservation and storage of to be freeze-fractured samples. (**a**) Copper profiles. (**b**) Metal pipe for storage of the propane/ethane mixture with an inner receptable at one end. (**c**) Arrangement of the dewar with the inner receptable in liquid nitrogen. (**d**) Storage container for frozen samples. (**e**) Arrangement of the storage container in a dewar filled with liquid nitrogen

3. Fill another dewar (dewar 2) with liquid nitrogen and equilibrate the cooling coil within it.
4. Pipeline the ethane/propane mixture through the cooling coil in the liquid nitrogen and collect the liquid ethane/propane mixture in the receptable in dewar 1 (Fig. 1c).

3.2 Cryopreservation

Do not prefix the specimen with any chemical fixatives (e.g., glutaraldehyde, paraformaldehyde)! The method is based on cryopreservation of the native state.

1. For in-solution samples.

   Liposomes [32], biochemical preparations of cellular organelles, preparations of membrane compartments, such as synaptosomes [36], suspensions of cells [15].

   (a) Prepare a highly concentrated suspension of cells, biochemical preparations of cellular membrane compartments, of isolated organelles, or of liposomes (*see* **Note 4**).

   (b) Enclose 1–2 μl of the concentrated suspension between two copper sandwich carriers by using forceps (*see* **Note 5**).

   (c) Rapidly freeze the specimen in liquid nitrogen-cooled liquid ethane/propane by plunging the sandwich into the liquid nitrogen-cooled liquid ethane/propane very quickly (Fig. 2a, b). This prevents ice formation. Release the sandwich into the ethane/propane.

   (d) Collect the frozen sandwiches in a numbered compartment inside a liquid nitrogen-filled dewar (dewar 3) until freeze-fracturing (Fig. 2c).

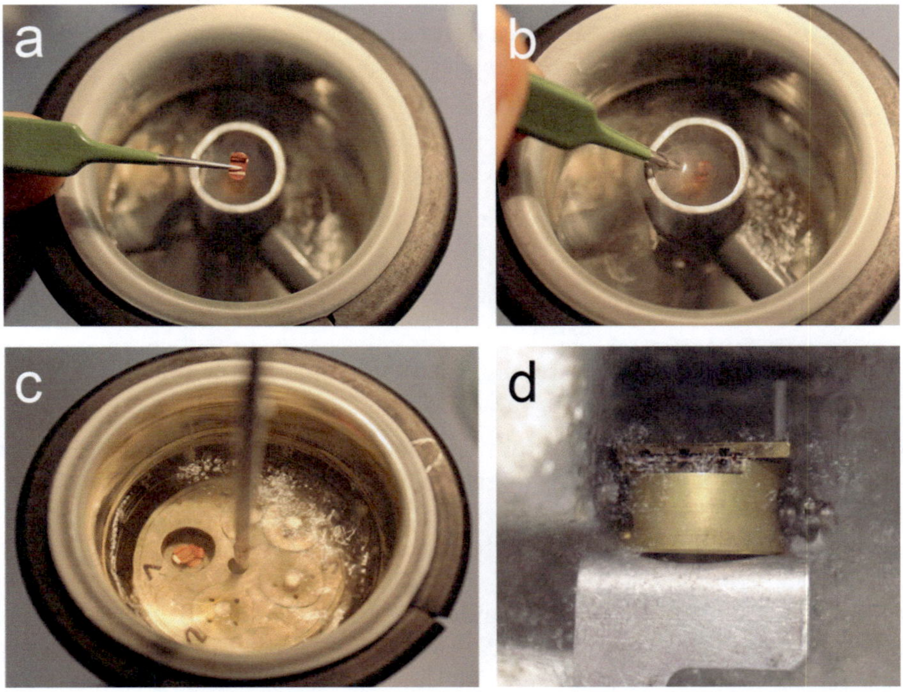

Fig. 2 Illustration of the plunge-freezing process of a sample. (**a**, **b**) Forceps are used to put the copper profiles together and to subsequently plunge them rapidly into the liquid propane/ethane mixture (**b**). (**c**) Storage container inserted into a liquid nitrogen-filled dewar that is used for storage until freeze-fracturing. (**d**) Equilibration of a double-sandwich replica holder (device for later freeze-fracturing) loaded with three double-sandwiches with specimen in liquid nitrogen

2. For adherent cells.

   Adherent cells grown on sapphire discs fitting the size of your double sandwich holder of your freeze-fracture unit (cell lines; primary cultures) [18, 37].

   (a) Grow cells on sapphire discs (use sapphire discs to ensure thermal and physical stability during sample preparation and freeze-fracturing) placed into a 24-well plate until they reach about 75% confluency. In case you work with postmitotic cells, seed them in a suitable density and let their morphology develop until the morphological and/or physiological stage you want to examine (*see* **Note 6**).

   (b) Prepare a 0.8 mm copper sandwich-carrier with 2 μl 20% w/v BSA (*see* **Note** 7).

   (c) Take the sapphire discs with the cells grown on them out of the 24-well cell culture plate using forceps.

   (d) Carefully put together a sapphire disc and a copper carrier in an orientation that places the cells toward the copper carrier by using forceps and then freeze the sandwich with the cells in the center rapidly in liquid nitrogen-cooled liquid ethane/propane as described above.

3. For tissue samples (we have e.g., done brain and different muscle tissues) [16, 18].

   (a) Prepare and dissect the tissue and rinse with PBS.

   (b) Cut the tissue by using a vibratome in slices of 50 μm thickness or less, if possible (*see* **Note 8**).

   (c) Prepare a 0.6 mm copper sandwich-carrier with 2 μl 20% BSA (*see* **Note** 7).

   (d) Cut out a tissue sample, which fits the size of the copper carrier.

   (e) Carefully place the tissue sample onto the copper carrier with the BSA.

   (f) Carefully put on the second copper carrier by using forceps to generate a sandwich with the sample in the center and freeze the sandwich rapidly in liquid nitrogen-cooled liquid ethane/propane as above.

3.3 Freeze–Fracturing and Replication

1. Mount the sandwiches under liquid nitrogen into a precooled double replica sandwich holder (Fig. 2d) and transfer the closed sandwich holder rapidly onto a cooled table in a BAF400T freeze-fracture unit (BAL-TEC, Liechtenstein).

2. Once the vacuum in the freeze-facture unit reaches 1×10^{-6} mbar and the temperature is at -140 °C, perform the fracture by opening the sandwich holder stage by using the microtome knife mechanism in the freeze-fracture unit.

3. Start the electron beam gun evaporation of 20 nm carbon immediately after the fracturing. Add the carbon in a perpendicular manner to obtain optimal support of all parts of the membrane (*see* **Note 9**).
4. Then start the evaporation of 2 nm platinum (and traces of carbon) at an angle of 35° for shadowing. Control the layer thickness by a thin-layer quartz crystal monitor. In contrast to Fujimoto (1995), we use reverse evaporation (first carbon, then platinum). This leads to more efficient immunolabeling results [34, 35] (*see* **Note 10**).
5. After the shadowing by C/Pt evaporation, raise the pressure in the freeze-fracture unit to atmospheric pressure and remove the sandwich holder from the freeze-fracture unit.
6. Thaw the copper profiles with the replicated samples on the top and very carefully transfer them into SDS buffer placed into a well of a ceramic spotting dish (Fig. 3a) (*see* **Note 11**).

3.4 SDS Treatment and Immunolabeling

1. Incubate the replica overnight at room temperature in SDS buffer under gentle shaking. Prevent that the spotting dishes are drying out (cover with glass plate) (Fig. 3b).
2. Wash the replica in PBS four times (10 min each) by placing them (using a loop) onto fresh buffer in the spotting dish to remove most of the SDS buffer (*see* **Note 12**). Then incubate them on LBB for 30 min.
3. For immunolabeling, place the replicas in a drop (100 μl on parafilm) (Fig. 3c) of a diluted solution of the primary antibodies (e.g., 1:50 of rabbit polyclonal anti-caveolin1, Santa Cruz sc-894) in LBB overnight at 4 °C (*see* **Note 13**).
4. Afterward wash the samples three times in LBB.
5. Place the replica in a drop (100 μl on parafilm) of secondary antibody solution (e.g., 1:50 goat anti-rabbit IgG coupled with electron dense 10 nm gold particles) in LBB) for 2 h at room temperature. Secondary antibodies coupled with 5, 15, or 20 nm gold particles can also easily be visualized by TEM (*see* **Note 14**).
6. Rinse the replica 3 times with PBS.
7. Fix them with the fixative for 10 min.
8. Wash the immunolabeled replica two times for 10 min with *aqua bidest.*
9. Finally pick them onto mesh grids (300–400 mesh), store dry (Fig. 3d) (*see* **Note 15**), and analyze the grids by TEM (Fig. 4 for image examples).

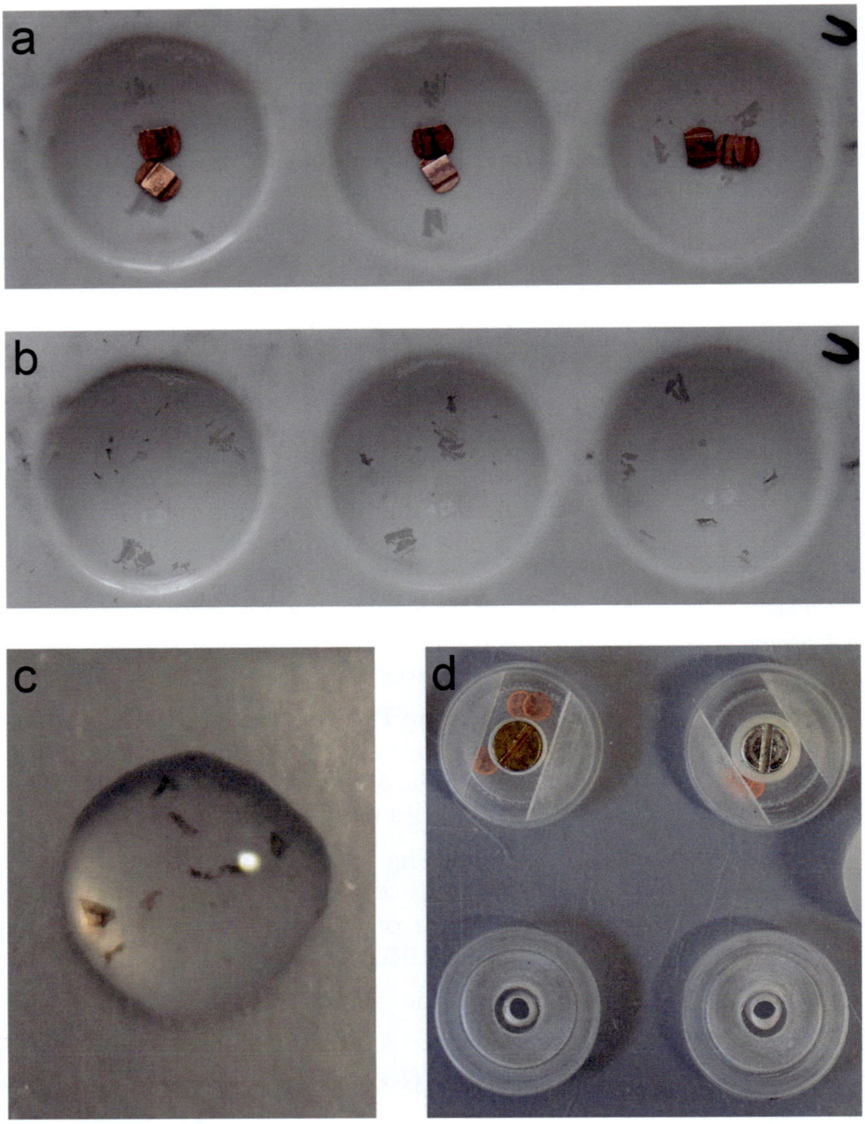

Fig. 3 Treatment and storage of replica after freeze-fracturing. (**a**) For SDS treatment prior to the immunolabeling carefully float the replica from the copper profiles onto SDS-buffer in a well of a ceramic spotting dish. (**b**) Replica in ceramic spotting dish after the copper profiles have been removed. (**c**) Replica on a 100 μl droplet of diluted antibody on parafilm (immunolabeling). (**d**) Self-made plate (dark gray surface) with four inserted shallow plastic containers for long-term grid storage. The two containers in the top row are closed by plastic lids held in place by a central screw and contain copper grids; the other containers are open to present the ring-shaped, shallow (approximately 2 mm deep) storage area and the central thread hole for the screw of the lid

3.5 Methodological Outlook

FRIL allows for qualitative analysis of the immunolabeled replica for the localizations of membrane-associated proteins at ultra-high resolution and for colocalizations of two (or maybe even three) membrane-associated proteins and/or specifically the membrane-associated subpools of proteins that can reversibly associate with the plasma membrane.

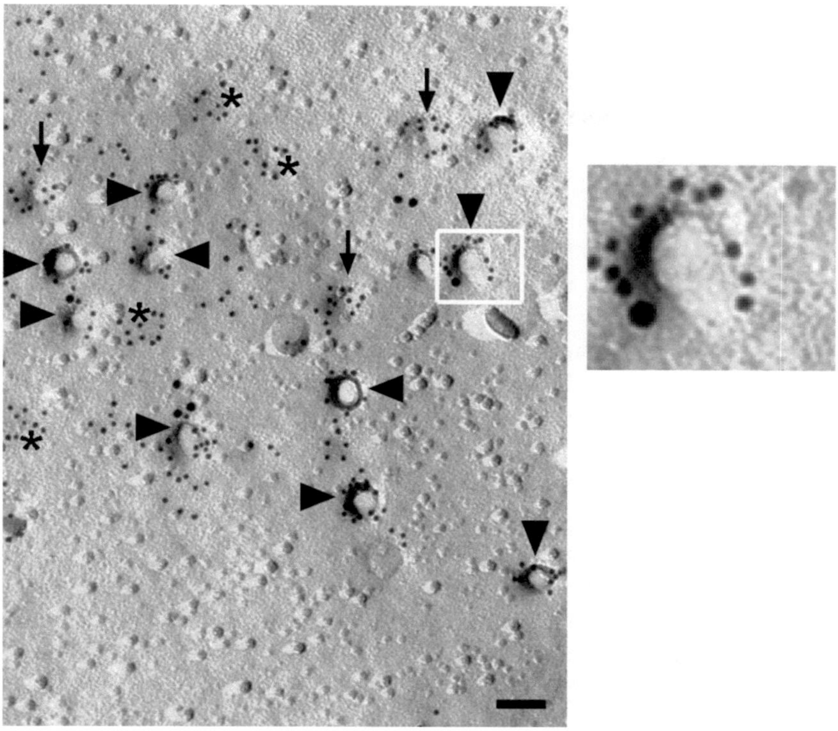

Fig. 4 Example of a TEM image of a freeze-fracture replica that was immunolabeled for caveolar components. TEM image of the cytosolic face of the inner leaflet of a cardiomyocyte isolated from a wild-type mouse that was immunolabeled for caveolin3 (the muscle specific isoform of the caveolin family) (10 nm gold) and the F-BAR protein syndapin III (15 nm gold). Note that deeply invaginated caveolae (arrowheads) and shallow caveolar structures (arrows) are visible that both are enriched for caveolin3 and that furthermore flat membrane areas can be found, which also show clusters of caveolin3 (asterisks). Right panel shows magnification of boxed area. Scale bar, 100 nm

FRIL also is particularly valuable for quantitative analyses of integral membrane proteins and partially membrane-inserted proteins, as the visualization by TEM provides easy access to the numbers of immunodetected molecules per membrane area of interest. The proteins can hereby be examined in regard of their spatial distribution to each other or in relation to colabeled further protein of interest provided that double labelings can be established. This way it e.g. becomes possible to study protein cluster formation in nanodomain areas over wide membrane areas [18, 37]). Since also membrane topology information is directly available at such wide membrane areas by using shadowing techniques, it is also straight forward to study the distribution of proteins or protein clusters in relation to specific membrane topologies, such as protrusions of different kinds [18, 37] or caveolae at different stages of invagination [15, 16]. These membrane topologies can also be examined further by specimen tilting in the electron beam and by EM tomography leading to highly detailed 3D representations of membranes decorated with the proteins of interest. We are sure that we will see a lot more of this type of analysis, as such studies will strongly promote our understanding of how membrane proteins are organized in the membrane and maybe also of how their organization is changed in response to different internal or external cues.

Conversely, the immunolabeling of membrane nanodomains or distinct features of membrane organization and topology is valuable to clearly identify their nature. One example for this was that shallow caveolar structures, i.e., membrane topologies that were not clearly recognizable as caveolin-coated structures by their profile in 3D, were identified as different stages of caveolar invagination by anti-caveolin immunolabelings [15, 16]. Even flat membrane areas can still be identified as distinct nanodomains, if they contain accumulations of a given protein of interest. This does apply not only to receptor fields [9, 10] or syndapin I nanodomains in the heads of dendritic spines of neurons [18] but also to flat caveolin-enriched areas of the plasma membrane [15, 16].

Furthermore, detailed three-dimensional reconstructions of membrane structures via tilt series in high resolution will increase our understanding of molecular mechanisms of cell biological processes by bridging the gap between structural and biochemical analyses on one side and visualizations of the process on the other side, as such views now are possible at ultrahigh resolution and in stunning detail. Closer analyzes of FRIL samples e.g. suggested that syndapin II and cavin1/PTRF have an asymmetric distribution at caveolar invaginations when viewed perpendicular [15]. First electron tomography examinations of invaginated membrane structures of cardiomyocytes labeled for both syndapin III and caveolin 3 with 15 nm and 10 nm gold, respectively, unveiled that the syndapin III distribution hereby was restricted to the edge of the caveolin coat outlined by anti-caveolin immunolabeling [16]. FRIL allows combinations of ultrahigh-resolution analyses of 3D membrane topology together with indubitable detections of components of interest at these distinct membrane profiles of interest - even at different stages of a given cell biological process or during a certain disease or physiological condition. Thus, this exciting technique provides deep insights into molecular processes at membranes and therefore opens up completely new horizons in membrane research.

4 Notes

1. You will need 10 ml SDS buffer per sample preparation. Thus, it is convenient to use 10 ml aliquots in plastic tubes with screw-cap.

2. To dissolve the BSA and teleostean gelatin it helps to warm up the PBS to 37 °C and to stir with a magnetic stir bar.
3. Use a fume hood while handling glutaraldehyde.
4. Use warm cell culture medium to wash and concentrate cells by centrifugation and also use warm buffers for liposomes in *in vitro* reconstitutions and for the analyses of membrane compartments and/or organelles, if possible.
5. It is important to suck off the excess of solution at the edge of the sandwich carrier by using a filter paper. This prevents ice formation.
6. If you use primary hippocampal neurons take a developmental delay of about 2 days into account when compared to normal culturing of neuronal cells.
7. Resolve BSA in *aqua bidest.*
8. You can store the slices in a 24-wellplate filled with PBS until freezing.
9. To prevent sample etching by a delayed evaporation, check the evaporation guns before freeze-fracture! Use 1.95 kV acceleration voltage and 93 mA emission current for carbon electron beam evaporator and 1.35 kV and 89 mA for the platinum/carbon source.
10. Note that reverse evaporation has an influence on the spatial resolution of the replica. Membrane particles were first covered with carbon and then with platinum, as the contrast-enhancing layer. Thus, keep the carbon layer thin.
11. To thaw the copper profiles, put the profiles using forceps shortly onto the back of your hand, then float the replica from the copper profiles by dipping them slowly (!) and using an angle of 45° into the SDS-buffer. Remove the copper profiles from the dish.
12. The handling of the replicas is done under a stereomicroscope using a loop made of platinum wire.
13. Prevent that the replica is curling. In case it happened that replica has furled, then use the platinum wire loop to lift the curled replica. Hold it for some seconds. Then carefully put it back onto the buffer surface. It should then flatten again. Repeat if necessary.
14. Note that smaller gold particles allow for better accessibility for epitopes. 5 nm gold or even nanogold tools therefore have advantages over, e.g., 20 nm gold. However, the detection of larger gold particles is much easier, as they offer a better contrast and signal-to-noise ratio when compared to the shadowed replica and will even be visible at the lower magnifications that are suitable for overviews over larger membrane

areas. Thus, in many cases, you will have to compromise between the extremes and/or optimize this parameter for your experiments. Starting with 10 nm or 15 nm gold may be advisable for most applications.

15. For storage we use self-made plastic containers with screw caps (Fig. 3d).

Acknowledgments

We would like to acknowledge the support from the Electron Microscopy Center, Jena University Hospital.

References

1. Steere RL (1957) Electron microscopy of structural detail in frozen biological specimens. J Biophys Biochem Cytol 3(1):45–60
2. Deamer DW, Branton D (1967) Fracture planes in an ice-bilayer model membrane system. Science 158(3801):655–657
3. Singer SJ, Nicolson GL (1972) The fluid mosaic model of the structure of cell membranes. Science 175(4023):720–731
4. Severs NJ (2007) Freeze-fracture electron microscopy. Nat Protoc 2(3):547–576
5. da Silva PP, Branton D (1970) Membrane splitting in freeze-etching covalently bound ferritin as a membrane marker. J Cell Biol 45 (3):598–605
6. Dinchuk JE, Johnson TJ, Rash JE (1987) Postreplication labeling of E-leaflet molecules: membrane immunoglobulins localized in sectioned, labeled replicas examined by TEM and HVEM. J Electron Microsc Tech 7(1):1–16
7. Gruijters WT, Kistler J, Bullivant S et al (1987) Immunolocalization of MP70 in lens fiber 16-17-nm intercellular junctions. J Cell Biol 104(3):565–572
8. Fujimoto K (1995) Freeze-fracture replica electron microscopy combined with SDS digestion for cytochemical labeling of integral membrane proteins. Application to the immunogold labeling of intercellular junctional complexes. J Cell Sci 108(11):3443–3449
9. Mansouri M, Kasugai Y, Fukazawa Y et al (2015) Distinct subsynaptic localization of type 1 metabotropic glutamate receptors at glutamatergic and GABA ergic synapses in the rodent cerebellar cortex. Eur J Neurosci 41 (2):157–167
10. Kasugai Y, Swinny JD, Roberts JDB et al (2010) Quantitative localisation of synaptic and extrasynaptic GABAA receptor subunits on hippocampal pyramidal cells by freeze-fracture replica immunolabelling. Eur J Neurosci 32(11):1868–1888
11. Indriati DW, Kamasawa N, Matsui K et al (2013) Quantitative localization of Cav2.1 (P/Q-type) voltage-dependent calcium channels in Purkinje cells: somatodendritic gradient and distinct somatic coclustering with calcium-activated potassium channels. J Neurosci 33 (8):3668–3678
12. Kaufmann W, Kasugai Y, Ferraguti F et al (2010) Two distinct pools of large-conductance calcium-activated potassium channels in the somatic plasma membrane of central principal neurons. Neuroscience 169 (3):974–986
13. Fujimoto T, Kogo H, Nomura R et al (2000) Isoforms of caveolin-1 and caveolar structure. J Cell Sci 113(19):3509–3517
14. Westermann M, Steiniger F, Richter W (2005) Belt-like localisation of caveolin in deep caveolae and its re-distribution after cholesterol depletion. Histochem Cell Biol 123 (6):613–620
15. Koch D, Westermann M, Kessels MM et al (2012) Ultrastructural freeze-fracture immunolabeling identifies plasma membrane-localized syndapin II as a crucial factor in shaping caveolae. Histochem Cell Biol 138 (2):215–230
16. Seemann E, Sun M, Krueger S et al (2017) Deciphering caveolar functions by syndapin III KO-mediated impairment of caveolar invagination. elife 6:e29854
17. Fujimoto T, Kogo H, Ishiguro K et al (2001) Caveolin-2 is targeted to lipid droplets, a new "membrane domain" in the cell. J Cell Biol 152(5):1079–1085

18. Schneider K, Seemann E, Liebmann L et al (2014) ProSAP1 and membrane nanodomain-associated syndapin I promote postsynapse formation and function. J Cell Biol 205(2):197–215
19. Qualmann B, Roos J, DiGregorio PJ et al (1999) Syndapin I, a synaptic dynamin-binding protein that associates with the neural Wiskott-Aldrich syndrome protein. Mol Biol Cell 10 (2):501–513
20. Qualmann B, Kelly RB (2000) Syndapin isoforms participate in receptor-mediated endocytosis and actin organization. J Cell Biol 148 (5):1047–1062
21. Modregger J, Ritter B, Witter B et al (2000) All three PACSIN isoforms bind to endocytic proteins and inhibit endocytosis. J Cell Sci 113 (24):4511–4521
22. Kessels MM, Qualmann B (2004) The syndapin protein family: linking membrane trafficking with the cytoskeleton. J Cell Sci 117 (Pt 15):3077–3086
23. Quan A, Robinson PJ (2013) Syndapin–a membrane remodelling and endocytic F-BAR protein. FEBS J 280(21):5198–5212
24. Kessels MM, Qualmann B (2015) Different functional modes of BAR domain proteins in formation and plasticity of mammalian postsynapses. J Cell Sci 128(17):3177–3185
25. McMahon HT, Gallop JL (2005) Membrane curvature and mechanisms of dynamic cell membrane remodelling. Nature 438 (7068):590–596
26. Frost A, Unger VM, De Camilli P (2009) The BAR domain superfamily: membrane-molding macromolecules. Cell 137(2):191–196
27. Qualmann B, Koch D, Kessels MM (2011) Let's go bananas: revisiting the endocytic BAR code. EMBO J 30(17):3501–3515
28. Peter BJ, Kent HM, Mills IG et al (2004) BAR domains as sensors of membrane curvature: the amphiphysin BAR structure. Science 303 (5657):495–499
29. Wang Q, Navarro MV, Peng G et al (2009) Molecular mechanism of membrane constriction and tubulation mediated by the F-BAR protein Pacsin/Syndapin. Proc Natl Acad Sci U S A 106(31):12700–12705
30. Monier S, Parton RG, Vogel F et al (1995) VIP21-caveolin, a membrane protein constituent of the caveolar coat, oligomerizes in vivo and in vitro. Mol Biol Cell 6(7):911–927
31. Parton RG, Hanzal-Bayer M, Hancock JF (2006) Biogenesis of caveolae: a structural model for caveolin-induced domain formation. J Cell Sci 119(5):787–796
32. Zobel T, Brinkmann K, Koch N et al (2015) Cooperative functions of the two F-BAR proteins Cip4 and Nostrin in regulating E-cadherin in epithelial morphogenesis. J Cell Sci 128(3):499–515
33. Schlörmann W, Steiniger F, Richter W et al (2010) The shape of caveolae is omega-like after glutaraldehyde fixation and cup-like after cryofixation. Histochem Cell Biol 133 (2):223–228
34. Fujita A, Fujimoto T (2007) Quantitative retention of membrane lipids in the freeze-fracture replica. Histochem Cell Biol 128 (5):385–389
35. Schlörmann W, John M, Steiniger F et al (2007) Improved antigen retrieval in freeze-fracture cytochemistry by evaporation of carbon as first replication layer. Histochem Cell Biol 127(6):633–639
36. Bocker HT, Heinrich T, Liebmann L et al (2019) The Na+/H+ exchanger Nhe1 modulates network excitability via GABA release. Cereb Cortex 29(10):4263–4276
37. Wolf D, Hofbrucker-MacKenzie SA, Izadi M, Seemann E, Steiniger F, Schwintzer L, Koch D, Kessels MM, Qualmann B (2018) Ankyrin repeat-containing N-Ank proteins shape cellular membranes. Nat Cell Biol 21 (10):1191–1205

Chapter 3

Analysis of Caveolin in Primary Cilia

Stine K. Morthorst, Johanne B. Mogensen, Søren T. Christensen, and Lotte B. Pedersen

Abstract

Recent evidence has indicated that caveolins are localized at the base of primary cilia, which are microtubule-based sensory organelles present on the cell surface, and that Caveolin-1 (CAV1) plays important roles in regulating ciliary membrane composition and function. Here we describe methods to analyze the localization and function of CAV1 in primary cilia of cultured mammalian cells. These include methods for culturing and transfecting mammalian cells with a CAV1-encoding plasmid or small interfering RNA (siRNA), analysis of mammalian cells by immunofluorescence microscopy (IFM) with antibodies against ciliary markers and CAV1, as well as methods for analyzing ciliary CAV1 function in siRNA-treated cells by IFM and cell-based signaling assays.

Key words Mammalian cell culture, Primary cilia, Caveolin, Immunofluorescence microscopy, siRNA-mediated knockdown, qPCR

1 Introduction

Primary cilia are dynamic, microtubule-based sensory organelles that are present in a single copy on the surface of many different vertebrate cell types, including most cells of the human body. They play a critical role in regulating a variety of signaling pathways during development and tissue homeostasis, including Sonic hedgehog (SHH) [1], G-protein-coupled receptor [2], TGFβ/BMP and receptor tyrosine kinase (RTK) signaling [3], and ciliary dysfunction has been linked to a growing number of pleiotropic diseases called ciliopathies [4]. The core of the cilium, the microtubule-based axoneme, is covered by a bilayer lipid membrane that is continuous with the plasma membrane of the cell body, but enriched in specific proteins and lipids that endow the organelle with unique signaling properties (Fig. 1). For example, recent studies showed that the membrane of primary cilia contains

Stine K. Morthorst and Johanne B. Mogensen are Co-first authors.

Cedric M. Blouin (ed.), *Caveolae: Methods and Protocols*, Methods in Molecular Biology, vol. 2169,
https://doi.org/10.1007/978-1-0716-0732-9_3,

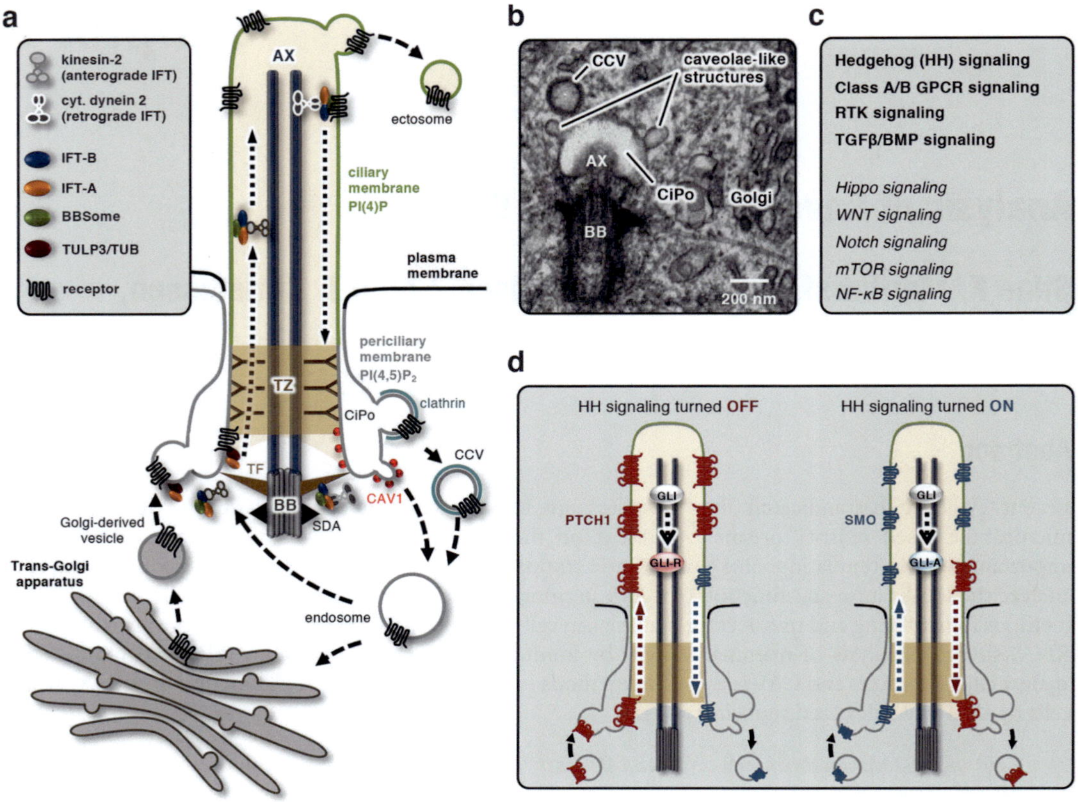

Fig. 1 (**a**) Simplified diagram of a primary cilium with associated trafficking processes. Modified from [32]. Abbreviations: AX, axoneme; BB, basal body; BBSome, Bardet Biedl Syndrome protein complex; CCV, clathrin-coated vesicle; CiPo, ciliary pocket; IFT, intraflagellar transport; RTK, receptor tyrosine kinase; SDA, subdistal appendage; TF, transition fiber; TULP3, Tubby (TUB)-like protein 3; TZ, transition zone. (**b**) Transmission electron micrograph of a thin-sectioned human foreskin fibroblast showing a longitudinal section of the base of a primary cilium (arrowhead) emanating from the CiPo. Reproduced from [21] with permission. (**c**) Overview of some of the main signaling pathways known to be associated with primary cilia. (**d**) Simplified schematic showing coordination of Hedgehog (HH) signaling by primary cilia. GLI-A, GLI activator; GLI-R, GLI repressor

higher levels of phosphatidylinositol 4-phosphate (PI4P) than the adjacent periciliary membrane and that ciliary enrichment of PI4P is critical for controlling SHH signaling output [5, 6]. Ciliary membrane compartmentalization is thus vital for signaling and relies on the transition zone located between the basal body and cilium proper [7], as well as on microtubule-based transport mechanisms such as intraflagellar transport (IFT) of ciliary cargo within the organelle [8] and cytoplasmic dynein-mediated transport of vesicles and membrane receptors to the ciliary base [9]. In addition, ciliary content may be influenced by ectocytosis of vesicles from the ciliary membrane [9–12] and by clathrin- and caveolin-mediated endocytic events at the periciliary membrane or pocket region (Fig. 1) [13–15] (*see* **Note 1**). For example, a recent study in

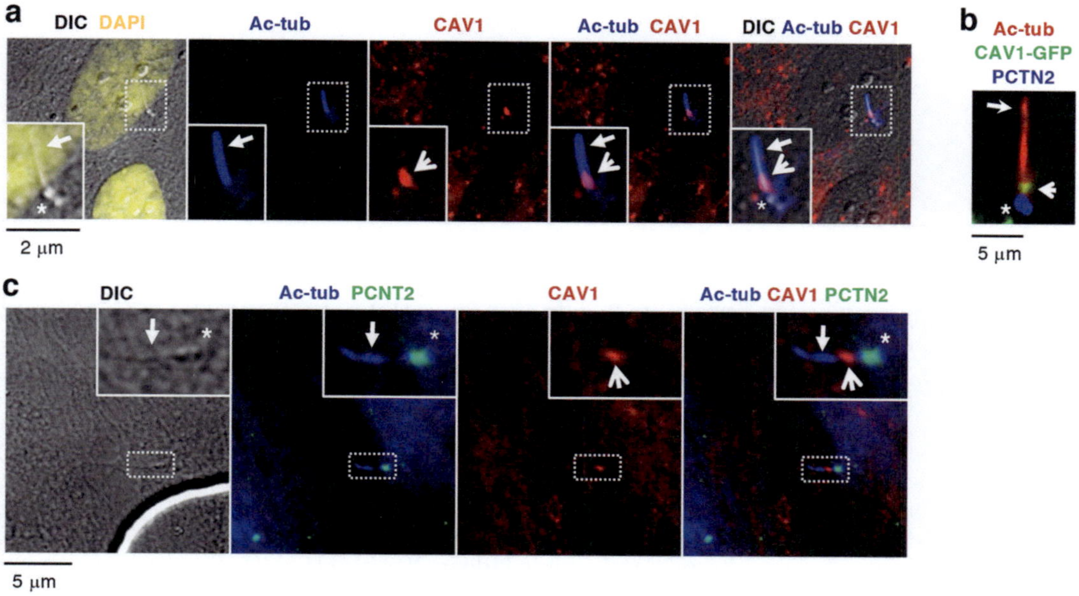

Fig. 2 Immunofluorescence micrographs showing localization of endogenous (**a, c**) and GFP-tagged CAV1 (CAV1-GFP) (**b**) at the primary cilium in PFA-fixed hTERT-RPE1 cells (**a, b**) and mouse embryonic fibroblasts (**c**), respectively. Open arrows point to CAV1 localizing distal to the ciliary basal body/centrosome marked by PCTN2 (asterisk). The cilium is indicated with a closed arrow. DIC: digital interference contrast. Modified from [17] with permission

Caenorhabditis elegans implicated Caveolin-1 (CAV1) in endocytosis at the periciliary membrane and regulation of ciliary localization and function of polycystins in male-specific sensory neurons [16], whereas we showed enrichment of CAV1 at the base of primary cilia in cultured hTERT-immortalized human retinal pigment epithelial (hTERT-RPE1) cells and mouse embryonic fibroblasts (Fig. 2; *see* **Note 2**), which is important for regulating cilia-mediated SHH signaling [17]. Cilia-related functions for caveolins are also supported by immunolocalization studies of rodent olfactory neurons and spermatozoa [18–20], and by electron microscopic observations of structures resembling caveolae at the periciliary membrane/pocket in cultured human foreskin fibroblasts (Fig. 1) [15, 21]. Although current evidence points to a role for caveolins in regulating ciliary membrane composition and signaling, presumably in a cell-type-dependent manner [16, 17], the precise function (s) of caveolins at cilia remains poorly understood.

Here we describe methods to analyze the localization and function of CAV1 in primary cilia of cultured mammalian cells. These include methods for culturing and transfecting mammalian cells with a CAV1-encoding plasmid or small interfering RNA (siRNA), analysis of mammalian cells by immunofluorescence microscopy (IFM) with antibodies against ciliary markers and CAV1, as well as methods for analyzing ciliary CAV1 function in siRNA-treated cells by IFM and cell-based SHH-signaling assays [17]. The methods described are optimized for the hTERT-RPE1 cell line (*see* **Note 3**), but are also applicable, with minor modifications, to other CAV1-expressing mammalian cell types such as mouse fibroblasts [17].

2 Materials

2.1 Mammalian Cell Culture and Transfection

1. Plasmids (*see* **Note 4**): CAV1-mEGFP/FRT/TO [22] and ShhN expression construct coding for the amino-terminal fragment (amino acids 1–201) of rat Shh [23].
2. siRNA: control siRNAi (5′-UAAUGU AUUGGAAUGCAUA (dTdT)-3′) and human CAV1-specific siRNA (5′-AAGAGC UUCCUGAUUGAGA-3′) [24].
3. Mammalian cell lines: human hTERT-RPE1 cells (*see* **Note 3**); human embryonic kidney 293 T (HEK293T) cells from American Type Culture Collection (ATCC) (CRL-3216) for ShhN expression.
4. Plasticware: T75 and T175 flasks, 6-, 10-, and 15-cm diameter Petri dishes, 6-well plates for cell culturing. Falcon 50-ml conical centrifuge tubes, 50-ml sterile luer lock syringes, and sterile 0.2-μM syringe disc-type filters for filtration of ShhN-conditioned medium.
5. Growth medium: Dulbecco's modified Eagle's medium (DMEM) with penicillin/streptomycin (p/s; 100 U/ml) and 10% heat-inactivated fetal bovine serum (FBS).
6. Starvation medium: DMEM.
7. Phosphate-buffered saline (PBS): 8 g NaCl, 0.2 g KCL, 1.15 g $NaHPO_4$, 0.2 g KH_2PO_4, ddH_2O to 1 l. Adjust pH to 7.4 with NaOH solution.
8. Trypsin mix: 1% trypsin-EDTA in PBS.
9. Transfection reagents: FuGENE®6 (Promega) and DharmaFECT Duo (Dharmacon).

2.2 Immunofluorescence Microscopy Analysis

1. Microscope slides and glass coverslips 12 mm in diameter.
2. PBS.
3. 4% Paraformaldehyde (PFA) in PBS.
4. Permeabilization buffer: 0.1% (v/v) Triton X-100 and 1% (v/v) bovine serum albumin (BSA) in PBS.
5. Blocking buffer: 2% (w/v) BSA in PBS.
6. Alternative blocking buffer: 0.2% (w/v) saponin in 2% (w/v) BSA in PBS.

7. Primary antibodies (dilutions in parentheses): mouse antiacetylated α-tubulin (Sigma, cat# T7451; 1:2000), rabbit anti-SMO (Abcam, cat# ab7817; 1:200), rabbit anti-CAV1 (Cell Signaling Technology, cat# 3238; 1:200), goat anti-PCTN2 (Santa Cruz, cat# sc-28,145; 1:500), rabbit anti-CEP164 (Sigma, cat# HPA037606; 1:500), and rabbit anti-ARL13B (ProteinTech cat# 17711–1-AP; 1:500).
8. Secondary antibodies (diluted 1:600): Alexa Flour 350-conjugated Donkey anti-mouse or Donkey anti-goat; Alexa Flour 488-conjugated Donkey anti-rabbit or Donkey anti-goat; Alexa Flour 568-conjugated Donkey anti-mouse, Donkey anti-rabbit, or Donkey anti-goat.
9. 20 mg/ml 4′, 6-diamidino-2-phenylindole, dihydrochloride (DAPI) solubilized in ddH_2O (should be aliquoted and stored at −20 °C in the dark).
10. Mounting reagent (5 ml): 4.5 ml glycerol, 0.5 ml 10× PBS, 0.1 g *N*-propyl gallate (*see* **Note 5**).
11. Humidity chamber (*see* **Note 6**).
12. Nail polish.

2.3 RNA Isolation and Reverse Transcription (RT)-Quantitative PCR (qPCR) Analysis

1. PBS.
2. RNase-free cell scrapers/rubber policemen.
3. Nonstick, RNase-free microcentrifuge tubes.
4. β-mercaptoethanol.
5. RNA extraction kit.
6. Reverse transcriptase.
7. dNTP-mix: 10 mM of each dNTP.
8. RNase inhibitor.
9. Reducing solution: 1 M dithiothreitol (DTT) solution.
10. DNase/RNase-free water.
11. 96-/348-well plates and adhesive films or PCR Strip Tubes with lids.
12. Kit for qPCR containing hot start PCR enzyme, dNTP mixture, Mg^{2+}, RNaseH, SYBR Green I and reference dye (50× solution).
13. Specific primers for qPCR (*see* Table 1) and random primers (oligodeoxyribonucleotides, mostly hexamers).

Table 1
Primers for qPCR analysis of human *GLI1*

Target gene	Sequence
GLI1	F: 5′ CAGGGAGTGCAGCCAATACAG 3′ R: 5′ GAGCGGCGGCTGACAGTATA 3′
B2M	F: 5′ GTTTACTCACGTCATCCAGCAGAGAATG 3′ R: 5′ CCATGTTTGATGTATCTGAGCAGGTTGC 3′

3 Methods

3.1 Transfection of hTERT-RPE1 Cells with CAV1-mEGFP/FRT/TO Plasmid for IFM Analysis

1. Culture hTERT-RPE1 cells in T75 flasks in 10 ml growth medium at 37 °C, 5% CO_2, 95% humidity.
2. Split cells when they reach 80–100% confluence. Wash the cells in 37 °C PBS and trypsinize with 2 ml trypsin-EDTA (1%) until the cells detach from the bottom of the flask.
3. Resuspend cells in 8 ml 37 °C fresh growth medium and transfer to a new T75 flask or prepare for IFM analysis as follows.
4. Transfer approximately 15% of the resuspended cells to a 6-well culture dish containing microscope glass coverslips (max 4 coverslips per well) and grow cells to 60–80% confluence at 37 °C, 5% CO_2, 95% humidity.
5. Prepare transfection mix immediately before transfection: in a sterile microcentrifuge tube, mix 100 ng plasmid with 3 μl Fugene 6 and 46 μl starvation medium. Incubate for 30 min at room temperature. During incubation, aspirate the growth medium in the 6-well culture dishes and add 1.5 ml starvation medium.
6. Slowly drip the transfection mix onto the cells grown on the coverslips in the 6-well culture dishes and incubate cells for 2 h at 37 °C, 5% CO_2, 95% humidity.
7. Remove starvation medium and transfection mix and wash gently in 37 °C PBS. Then add 1.5 ml fresh starvation medium and incubate cells for no longer than 16 h to induce growth arrest and ciliogenesis (*see* **Note 7**).

3.2 Production of ShhN-Conditioned Medium in HEK293T Cells

1. The ShhN-conditioned medium can be used to stimulate cultured mammalian cells to assess SHH pathway activity [23]. For the production of ShhN-conditioned medium, culture HEK293T cells in growth medium in a T175 flask and split cells when they are 80–90% confluent (*see* **Note 8**). Wash cells in 37 °C PBS and trypsinize in 4 ml trypsin-EDTA (1%) until the cells detach from the bottom of the flask.

2. Resuspend cells in 16 ml 37 °C fresh growth medium and transfer to a new T175 flask or prepare for experiments as follows.
3. For transfection with ShhN expression plasmid, seed HEK293T cells in 150-mm dishes and grow to 60–80% confluence at 37 °C, 5% CO_2, 95% humidity. A confluent T175 flask makes six 150-mm dishes.
4. Prepare transfection mixture immediately before transfection: in a sterile microcentrifuge tube, mix 8 μg plasmid with 28 μl Fugene6 and 966 μl starvation medium. Incubate for 30 min at room temperature. During incubation, aspirate the medium from the cells and add 19 ml of fresh starvation medium.
5. Slowly drip the transfection mixture onto the cells and incubate for 4 h at 37 °C, 5% CO_2, 95% humidity.
6. Aspirate transfection mixture and add 20 ml normal growth medium; incubate cells for 48 h at 37 °C, 5% CO_2, 95% humidity.
7. Aspirate conditioned medium and filter through a sterile 0.20-μM syringe filter into a 50-ml Falcon tube.
8. Store at −80 °C until use.

3.3 Transfection of hTERT-RPE1 Cells with CAV1-Specific siRNA for IFM Analysis

1. Cultivate and seed hTERT-RPE1 cells as in Subheading 3.1, **steps 1–4**.
2. Prepare three separate sterile microcentrifuge tubes containing the following:
   (a) Tube A: Mix 200 μl starvation medium with 10 μl Dharmafect Duo.
   (b) Tube B: Mix 100 μl starvation medium with CAV1-specific siRNA (final concentration: 250 nM).
   (c) Tube C: Mix 100 μl starvation medium with control siRNA (final concentration: 250 nM).
3. Incubate the tubes at room temperature for 5 min and then carefully mix the tubes by transferring 100 μl of tube A into tubes B and C; incubate for 30 min at room temperature.
4. During incubation, aspirate growth medium from hTERT-RPE cells and add 1.8 ml fresh starvation medium.
5. Carefully drip transfection mixture onto cells and incubate for 6 h at 37 °C, 5% CO_2, 95% humidity.
6. Change medium to normal growth medium and incubate for 18 h at 37 °C, 5% CO_2, 95% humidity.
7. Repeat transfection (**steps 2–6**; *see* **Note 9**).
8. Aspirate medium, add 3 ml of starvation medium and incubate for 24 h at 37 °C, 5% CO_2, 95% humidity prior to IFM analysis.

If the role of CAV1 in SHH signaling is to be investigated, the medium is changed to a 1:1 mixture of fresh starvation medium and negative control HEK293T-conditioned medium in one dish as a negative control (*see* **Note 8**) and a 1:1 mixture of starvation medium and ShhN-conditioned medium in another dish, and cells are incubated for additional 24 h at 37 °C, 5% CO_2, 95% humidity.

9. Use IFM (Fig. 3a) or western blot analysis with CAV1 antibody to assess the efficiency of CAV1 depletion (*see* **Note 9**).

3.4 Immunofluorescence Microscopy and Image Analysis of Ciliary Markers and CAV1 in hTERT-RPE1 Cells

1. Remove medium from culture dish where cells were grown on glass coverslips and gently wash the cells in ice-cold (4 °C) PBS.
2. Fix cells with 4% PFA solution for 15 min at room temperature directly in the cell-culture dish.
3. Gently wash cells twice with ice-cold PBS.
4. Permeabilize cells by incubating them in permeabilization buffer for 12 min. As an alternative to permeabilization with Triton X-100, the cells can be incubated with 0.2% saponin in blocking buffer for 1 h after PFA fixation. Importantly, the primary and secondary antibodies should then be diluted in blocking buffer with 0.2% saponin, and the subsequent washes must also be done using blocking buffer with 0.2% saponin.
5. Replace the permeabilization buffer with PBS and transfer coverslips to a humidity chamber (*see* **Note 6**) using forceps or scalpel with the cells facing up.
6. The following incubations with blocking buffer, antibody dilutions, and PBS are achieved by adding a 50–100 μl drop of liquid on top of each coverslip. Make sure the coverslips do not dry out.
7. As the coverslips are transferred to the humidity chamber, incubate with blocking buffer for 30 min.
8. Incubate with relevant primary antibodies diluted in blocking buffer (*see* Subheading 2.2, **step 7**) at room temperature for 1.5 h or at 4 °C overnight. For measurements of ciliary frequency (Fig. 3b) or length, primary antibodies against acetylated α-tubulin and/or the ciliary membrane marker ARL13B can be used. To monitor SHH pathway activity following stimulation with ShhN-conditioned medium, antibody against SMO can be used (Fig. 3c) (*see* **Note 10**).
9. Gently wash the coverslips 3 × 5 min with blocking buffer and incubate with appropriate secondary antibodies diluted in blocking buffer (*see* Subheading 2.2, **step 8**) for 45 min.
10. Wash the coverslips 5 min with blocking buffer, stain briefly (30 s) with DAPI diluted 1:5000 in PBS.

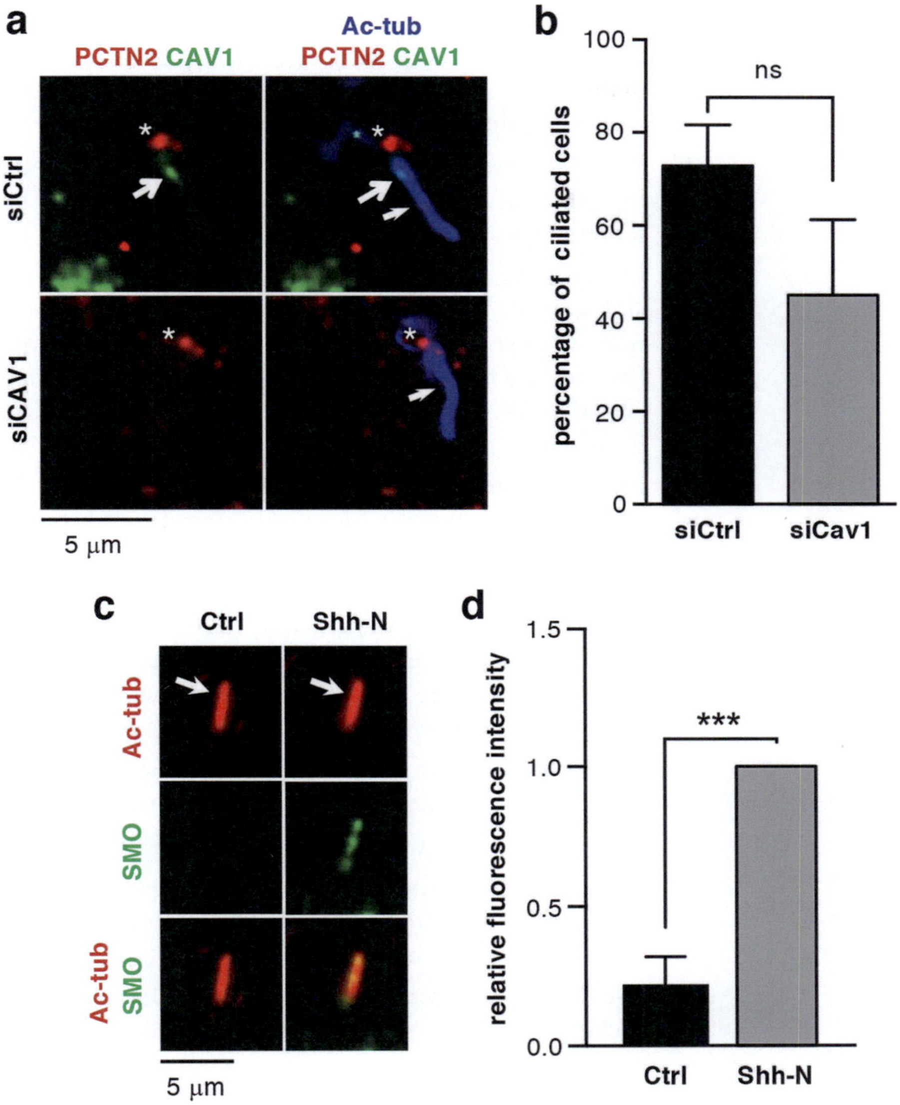

Fig. 3 Immunofluorescence micrographs of PFA-fixed hTERT-RPE1 cells illustrating the efficiency of siRNA-mediated depletion of CAV1 (**a**) and ciliary accumulation of SMO following treatment with ShhN-conditioned medium (**c**). The primary cilium is indicated with a closed arrow in (**a, c**). Open arrows in (**a**) point to CAV1 localizing distal to the ciliary basal body/centrosome (PCTN2; asterisk) of control siRNA-treated cells (siCtrl). (**b**) Quantification of the frequency of ciliated cells in control and CAV1-depleted cells indicates that CAV1 is not required for ciliogenesis in hTERT-RPE1 cells. At least 50 cells were analyzed per condition in three independent experiments; bars represent mean ± s.e.m.; P value (ns, $P \geq 0.05$, not significant) is derived from two-tailed *t*-test. (**d**) Quantification of relative fluorescence intensity of SMO staining in primary cilia based on images similar to those shown in (**c**). The relative fluorescence intensity of ciliary SMO was measured for at least 50 cells per condition ($n = 3$). Bars represent mean ± s.e.m. *P* value results from a two-tailed *t*-test ($***$, $P \leq 0.001$). Modified from [17] with permission

11. Gently wash the coverslips 3 × 5 min with PBS.
12. Mount the coverslip with cells facing down in a droplet of mounting reagent on a microscope slide cleaned in advance with 96% ethanol.
13. Remove excess mounting medium with a paper towel and seal the edges of the coverslips with nail polish. Store the microscope slides at 4 °C until imaging.
14. Use an epifluorescence microscope with an appropriate camera and software to visualize and capture images of the cells. We routinely use a fully motorized and automated Olympus IX83 Inverted microscope with a Hamamatsu ORCA-Flash 4.0 camera (C11440-22CU) and cellSens Dimension software from Olympus.
15. Process images for publication using image software. Examples of images showing localization of endogenous and GFP-tagged CAV1 at the base of primary cilia in hTERT-RPE1 cells are provided in Fig. 2a, b, respectively.
16. Use Olympus cellSens Dimension analysis software or other relevant software (e.g., ImageJ) for measurements of ciliary length. Draw a line along the ciliary axoneme using the Polyline function in the Count & Measure menu.
17. Similar software can be used for measurement of relative ciliary levels of SMO, as described in the following.
18. First, capture images using appropriate and similar settings.
19. Draw an outline around each cilium and use the Measurement and Region of Interest (ROI) function to measure the relevant mean fluorescence intensity in this region.
20. Use the same procedure to make a mean fluorescence intensity background reading of an area in the cytosol.
21. Subtract the measured background mean fluorescence intensity from the mean ciliary fluorescence intensity to obtain the relative fluorescence intensity of SMO in the cilium.
22. Use statistics software to perform a two-tailed *t*-test of at least three independent experiments.

3.5 RNA Isolation, cDNA Synthesis, and RT-qPCR Analysis to Measure SHH Pathway Activity

In addition to assessing ShhN responsiveness by quantifying relative ciliary SMO levels by IFM (*see* Subheading 3.4), measurement of relative *GLI1* mRNA levels by RT-qPCR can be used as a readout of pathway activity (*see* **Note 10**). In the following, we describe procedures for such an analysis in hTERT-RPE1 cells, which might be useful for confirming reported roles of CAV1 in SHH signaling [17, 25].

RNA Isolation:

1. Culture hTERT-RPE1 cells as in Subheading 3.1, **steps 1–3**, and seed cells in two 100-mm sterile Petri dishes.
2. When cells are confluent, aspirate growth medium and add 10-ml starvation medium; incubate cells for 24 h at 37 °C, 5% CO_2, 95% humidity to induce ciliogenesis.
3. Change medium to a 1:1 mixture of fresh starvation medium and negative control HEK293T-conditioned medium in one dish as a negative control (*see* **Note 8**) and add a 1:1 mixture of starvation medium and ShhN-conditioned medium in the second dish and incubate for 24 h at 37 °C, 5% CO_2, 95% humidity.
4. For RNA extraction, use a commercially available RNA extraction kit: For each dish, prepare a lysis buffer by mixing lysis buffer with a reducing agent to denature RNases (usually β-mercaptoethanol).
5. The following steps are done on ice. Aspirate medium and wash cells using ice-cold PBS.
6. Add lysis buffer to cells, scrape off the lysate using a sterile rubber policeman and transfer to a microcentrifuge tube.
7. Proceed with RNA extraction and purification by following the instructions provided by the manufacturer of the RNA extraction kit.
8. Elute RNA using 40-μl RNase-free water and store at −80 °C until use.

cDNA Synthesis:

9. Calculate RNA concentration (*C*) by measuring the absorbance at 260 nm (*A260*) in a UV spectrophotometer: *C* (mg/ml) = *A260*:0.025 (*see* **Note 11**).
10. For each sample, mix 250 ng random primers, dNTP-mix (10 mM), 1 μg RNA, and up to 13.5 μl nuclease-free water. Incubate 5 min at 65 °C (e.g., on a heating block or in a thermal cycler).
11. Put samples on ice for 1 min and spin down shortly to collect condensation.
12. Add 4 μl 5× First Strand Buffer, 1 μl 1 M DTT solution, 1 μl RNase inhibitor, and 0.5 μl reverse transcriptase (200 U/μl), so the total volume in each tube is 20 μl. Carefully mix by resuspending the samples and incubate for 5 min at room temperature.
13. Incubate in a thermal cycler at 60 °C for 50 min followed by 70 °C for 15 min.
14. Store cDNA at −20 °C.

qPCR Analysis:

15. Prepare a master mix that per sample contains 10 μl SYBR Green I, 2 μl forward primer (from 200 nM stock), 2 μl reverse primer (from 200 nM stock), 0.4 μl reference dye 50×, and add ddH_2O up to 19 μl. For each primer set, in this case specific for human *GLI1* and *B2M* (Table 1), a standard curve is prepared by 5× fivefold serial dilutions of cDNA of the untreated control sample.
16. Dispense 19 μl master mix into each reaction well corresponding to the number needed to run duplicates of the serial dilutions, triplicates of each sample of interest, and duplicates for negative controls containing ddH_2O instead of cDNA for each primer pair (*see* **Note 12**).
17. Pipette 1 μl cDNA sample into the appropriate wells (*see* **Note 13**) and seal the plate with adhesive film or close the PCR strip tubes using lids and centrifuge for 30 s at 500 rpm.
18. Run qPCR on a light cycler. The settings should be adjusted, based on the volume of master mix and the primers in the reactions. Our results [17] were obtained using the following program: 95 °C for 10 min, 40 cycles of (95 °C for 30 s, 64 °C for 1 min, 72 °C for 30 s) and 95 °C for 1 min.

4 Notes

1. The ciliary pocket is an invagination of the periciliary membrane found in some cell types such as mammalian fibroblasts and retinal pigment epithelial cells. It is a cellular hotspot for endocytosis and exocytosis of vesicles derived from or destined to the ciliary compartment [26].
2. In ref. 17, we reported that CAV1 is concentrated at the ciliary transition zone in hTERT-RPE1 and mouse embryonic fibroblasts, based on IFM analysis of ciliated cells with antibodies against the ciliary axoneme (acetylated α-tubulin) and basal body (PCTN2) markers. However, because the periciliary membrane in these cell types is often invaginated to form a pocket-like structure that surrounds the proximal part of the cilium (*see* **Note 1**, Fig. 1 and ref. 26), it cannot formally be excluded that the CAV1 localization observed at the ciliary base in these cells [17] reflects CAV1 presence in the periciliary/pocket membrane juxtaposed to the transition zone. High-resolution imaging approaches are needed to distinguish between these two possible sites of CAV1 localization at the ciliary base.

3. The hTERT-RPE1 cell line can be obtained from American Type Culture Collection (ATCC). This cell line is a diploid, human immortalized and adherent retinal epithelial-like cell line, which was derived from the RPE-340 cell line via transfection with the plasmid pGRN145 hTERT that can be obtained from ATCC. Cultures of hTERT-RPE1 cells ciliate efficiently when grown to confluence and/or upon deprivation of serum. The procedures described here for IFM analysis of CAV1 in ciliated hTERT-RPE1 cells can also be used for other cell types, such as mouse fibroblasts, which display prominent ciliary base accumulation of CAV1 similar to the hTERT-RPE1 cells [17]. Although hTERT-RPE1 cells were shown in several studies to respond to SHH ligand or agonists such as SMO agonist (SAG), e.g., by accumulating SMO in the primary cilium [17, 27, 28], mouse fibroblasts are frequently used for studies of primary cilia and SHH signaling [29] and generally respond better to SHH ligand and agonists compared to hTERT-RPE1 cells.
4. Plasmids can be prepared from *Escherichia coli* by standard approaches using a commercially available kit.
5. The *N*-propyl gallate solution can be stored several days at 4 °C.
6. A humidity chamber can be prepared by simply placing a moist piece of filter paper in a plastic dish. Add a piece of parafilm on top of the filter paper and place the coverslips on the parafilm, cells facing up. Cover the plastic dish with a lid.
7. To avoid plasmid overexpression artifacts, it may be necessary to optimize transfection and incubation times.
8. A negative control should be prepared simultaneously with the ShhN-conditioned medium. The negative control can be prepared and harvested by following the same protocol as for the ShhN-conditioned medium and either by transfecting with an empty vector, or simply by skipping the transfection step.
9. Repetition of transfection can be avoided if initial CAV1 knockdown efficiency is high.
10. In vertebrates, the SHH pathway requires the primary cilium [30]. In the off-state, the SHH receptor PTCH1 is enriched in the cilium and ciliary levels of SMO are low. Upon SHH stimulation, PTCH1 binds to SHH and exits the cilium concomitantly with ciliary accumulation of SMO [29]. This activates a series of downstream signaling events at the cilium, which culminate in the stabilization of the full-length, activator forms of the GLI transcription factors (GLI-A) and upregulated expression of SHH target genes such as *GLI1* and *PTCH1* [1] (Fig. 1d). Assessment of SHH pathway activity can therefore be monitored by, e.g., analyzing relative ciliary

levels of SMO by IFM, measuring cellular ratios of full-length versus truncated GLI protein by western blotting, and measurement of *GLI1* and *PTCH1* mRNA levels by RT-qPCR [17, 29, 31].

11. It is recommended to assess the quality of the RNA, e.g., by Northern blotting.
12. It is very important that the pipetting is accurate at this step.
13. Since nucleic acids are highly charged and stick to the walls of plastic tubes, we recommend to use Nonstick tubes. Furthermore, before pipetting cDNA into the qPCR wells, make sure to mix the samples by vortexing and then spin down using a microcentrifuge before transferring the sample. To be as accurate as possible, tips should be changed after every transfer.

Acknowledgments

The authors acknowledge funding from the Independent Research Fund Denmark (6108-00457B and 8020-00162B), the Novo Nordisk Foundation (NNF14OC0011535, NNF15OC0016886 and NNF18OC0053024), Brødrene Hartmanns Fond (A31662), Kræftens Bekæmpelse (R146-A9590-16-S2), Carlsberg Foundation (CF18-0294), and the University of Copenhagen Excellence Programme for Interdisciplinary Research (2016 Funds). JBM and SKM were partially supported by PhD fellowships from the Department of Biology, University of Copenhagen.

References

1. Bangs F, Anderson KV (2017) Primary cilia and mammalian hedgehog signaling. Cold Spring Harb Perspect Biol 9:a028175
2. Schou KB, Pedersen LB, Christensen ST (2015) Ins and outs of GPCR signaling in primary cilia. EMBO Rep 16:1099–1113
3. Christensen ST, Morthorst SK, Mogensen JB, Pedersen LB (2017) Primary cilia and coordination of receptor tyrosine kinase (RTK) and transforming growth factor beta (TGF-beta) signaling. Cold Spring Harb Perspect Biol 9: a028167
4. Reiter JF, Leroux MR (2017) Genes and molecular pathways underpinning ciliopathies. Nat Rev Mol Cell Biol 18:533–547
5. Garcia-Gonzalo FR, Phua SC, Roberson EC, Garcia G 3rd, Abedin M, Schurmans S, Inoue T, Reiter JF (2015) Phosphoinositides regulate ciliary protein trafficking to modulate hedgehog signaling. Dev Cell 34:400–409
6. Chavez M, Ena S, Van Sande J, de Kerchove d'Exaerde A, Schurmans S, Schiffmann SN (2015) Modulation of ciliary phosphoinositide content regulates trafficking and sonic hedgehog signaling output. Dev Cell 34:338–350
7. Garcia-Gonzalo FR, Reiter JF (2016) Open sesame: how transition fibers and the transition zone control ciliary composition. Cold Spring Harb Perspect Biol 9:a028134
8. Taschner M, Lorentzen E (2016) The Intraflagellar transport machinery. Cold Spring Harb Perspect Biol 8:a028092
9. Cao M, Ning J, Hernandez-Lara CI, Belzile O, Wang Q, Dutcher SK, Liu Y, Snell WJ (2015) Uni-directional ciliary membrane protein trafficking by a cytoplasmic retrograde IFT motor and ciliary ectosome shedding. eLife 4:e05242
10. Wood CR, Huang K, Diener DR, Rosenbaum JL (2013) The cilium secretes bioactive ectosomes. Curr Biol 23:906–911

11. Wood CR, Rosenbaum JL (2015) Ciliary ectosomes: transmissions from the cell's antenna. Trends Cell Biol 25:276–285
12. Nager AR, Goldstein JS, Herranz-Perez V, Portran D, Ye F, Garcia-Verdugo JM, Nachury MV (2017) An actin network dispatches ciliary GPCRs into extracellular vesicles to modulate signaling. Cell 168:252–263
13. Benmerah A (2013) The ciliary pocket. Curr Opin Cell Biol 25:78–84
14. Clement CA, Ajbro KD, de Jesus MPR H, Koefoed K, Vestergaard ML, Veland IR, Pedersen LB, Benmerah A, Andersen CY, Larsen LA, Christensen ST (2013) Regulation of TGFβ signaling by endocytosis at the pocket region of the primary cilium. Cell Rep 3:1806–1814
15. Pedersen LB, Mogensen JB, Christensen ST (2016) Endocytic control of cellular signaling at the primary cilium. Trends Biochem Sci 41:784–797
16. Scheidel N, Kennedy J, Blacque OE (2018) Endosome maturation factors Rabenosyn-5/VPS45 and Caveolin-1 regulate ciliary membrane and Polycystin-2 homeostasis. EMBO J 37:e98248
17. Schou KB, Mogensen JB, Nielsen BS, Morthorst SK, Aleliunaite A, Serra-Marques AMA, Saunier S, Bizet A, Veland IR, Akhmanova A, Christensen ST, Pedersen LB (2017) KIF13B establishes a CAV1-enriched microdomain at the ciliary transition zone to promote sonic hedgehog signaling. Nat Commun 8:14177
18. Schreiber S, Fleischer J, Breer H, Boekhoff I (2000) A possible role for caveolin as a signaling organizer in olfactory sensory membranes. J Biol Chem 275:24115–24123
19. Travis AJ, Merdiushev T, Vargas LA, Jones BH, Purdon MA, Nipper RW, Galatioto J, Moss SB, Hunnicutt GR, Kopf GS (2001) Expression and localization of caveolin-1, and the presence of membrane rafts, in mouse and Guinea pig spermatozoa. Dev Biol 240:599–610
20. Trevino CL, Serrano CJ, Beltran C, Felix R, Darszon A (2001) Identification of mouse trp homologs and lipid rafts from spermatogenic cells and sperm. FEBS Lett 509:119–125
21. Schrøder JM, Larsen J, Komarova Y, Akhmanova A, Thorsteinsson RI, Grigoriev I, Manguso R, Christensen ST, Pedersen SF, Geimer S, Pedersen LB (2011) EB1 and EB3 promote cilia biogenesis by several centrosome-related mechanisms. J Cell Sci 124:2539–2551
22. Hayer A, Stoeber M, Ritz D, Engel S, Meyer HH, Helenius A (2010) Caveolin-1 is ubiquitinated and targeted to intralumenal vesicles in endolysosomes for degradation. J Cell Biol 191:615–629
23. Zeng X, Goetz JA, Suber LM, Scott WJ Jr, Schreiner CM, Robbins DJ (2001) A freely diffusible form of sonic hedgehog mediates long-range signalling. Nature 411:716–720
24. Millan J, Hewlett L, Glyn M, Toomre D, Clark P, Ridley AJ (2006) Lymphocyte transcellular migration occurs through recruitment of endothelial ICAM-1 to caveola- and F-actin-rich domains. Nat Cell Biol 8:113–123
25. Yue S, Tang LY, Tang Y, Tang Y, Shen QH, Ding J, Chen Y, Zhang Z, Yu TT, Zhang YE, Cheng SY (2014) Requirement of Smurf-mediated endocytosis of Patched1 in sonic hedgehog signal reception. eLife 3:e02555
26. Ghossoub R, Molla-Herman A, Bastin P, Benmerah A (2011) The ciliary pocket: a once-forgotten membrane domain at the base of cilia. Biol Cell 103:131–144
27. Hirano T, Katoh Y, Nakayama K (2017) Intraflagellar transport-A complex mediates ciliary entry and retrograde trafficking of ciliary G protein-coupled receptors. Mol Biol Cell 28:429–439
28. Vuolo L, Stevenson NL, Heesom KJ, Stephens DJ (2018) Dynein-2 intermediate chains play crucial but distinct roles in primary cilia formation and function. eLife 7:e39655
29. Rohatgi R, Milenkovic L, Scott MP (2007) Patched1 regulates hedgehog signaling at the primary cilium. Science 317:372–376
30. Huangfu D, Liu A, Rakeman AS, Murcia NS, Niswander L, Anderson KV (2003) Hedgehog signalling in the mouse requires intraflagellar transport proteins. Nature 426:83–87
31. Eguether T, San Agustin JT, Keady BT, Jonassen JA, Liang Y, Francis R, Tobita K, Johnson CA, Abdelhamed ZA, Lo CW, Pazour GJ (2014) IFT27 links the BBSome to IFT for maintenance of the ciliary signaling compartment. Dev Cell 31:279–290
32. Morthorst SK, Christensen ST, Pedersen LB (2018) Regulation of ciliary membrane protein trafficking and signalling by kinesin motor proteins. FEBS J 285:4535–4564

Chapter 4

Method for Efficient Observation of Caveolin-1 in Plasma Membrane by Microscopy Imaging Analysis

Tomoya Yamaguchi, Toyoshi Fujimoto, and Takashi Takahashi

Abstract

Fluorescence microscopy is currently one of the more powerful and versatile techniques available for biological studies. With conventional biological immunofluorescence microscopy, caveolin-1 (CAV1) is visualized as numerous small dots, which are often distributed as a linear array or along the edge of the cell. Although its presence, as well as that of other proteins, can be detected by conventional immunofluorescence microscopy, those results do not clarify whether two different proteins exist in the plasma membrane of a specimen or how they are distributed two-dimensionally. Here, we describe an unroofing procedure that clearly reveals CAV1 localization in a single plane of the plasma membrane and also demonstrate a super-resolution structured illumination microscopy technique for observation of CAV1 in the plasma membrane.

Key words Immunofluorescence staining, Unroofing method, Nitrocellulose membrane, Plasma membrane, Super-resolution, Structured illumination microscopy

1 Introduction

Caveolae are small invaginations of the plasma membrane generally found in adipocytes, endothelial cells, and smooth muscle cells. Therefore, caveolae exist in a broad variety of cell types [1–3]. Numerous immunofluorescence studies have documented that a variety of signaling molecules are concentrated in caveolin-enriched fractions [2, 4]. Immunofluorescence labeling with the anti-CAV1 antibody reveals small dots in different types of cultured cells. In fibroblasts, a large number of dots are seen scattered in the outspread portion of the cells, with a frequent concentration in patches seen at the edge of the cell and in linear arrays over the cell surface [5]. Double immunofluorescence labeling has been used to show a general distributional similarity between CAV1 and a protein of interest. However, except for high-resolution micrography, in which each caveola can be observed as a punctate, it is not

Cedric M. Blouin (ed.), *Caveolae: Methods and Protocols*, Methods in Molecular Biology, vol. 2169,
https://doi.org/10.1007/978-1-0716-0732-9_4,

possible to determine whether the protein is in the caveolae or their vicinity.

Localization of labeling for CAV1 in the plasma membrane is determined using unroofed cultured cells [5–8]. For this method, a nitrocellulose filter is adhered to the dorsal surface of cultured cells and then used to pull it apart. When successfully performed, the dorsal plasma membrane and most of the cytoplasm are removed, with only the basal plasma membrane and a small number of cytoskeletal components remaining on the substrate. We previously demonstrated that colocalization of punctate signals of CAV1, an essential structural protein of caveolae, with those of ROR1, a protein required for sustaining caveolae structure as well as EGFR survival signaling pathway in its kinase-independent manner, can be verified in lung cancer cells by double immunofluorescence staining combined with an unroofing method (Fig. 1) [9, 10].

Structured illumination microscopy (SIM) is a method used to increase the spatial resolution of wide-field fluorescence microscopy beyond its classical limits by use of spatially structured illumination [11–13]. We also previously found that colocalization of CAV1 and ROR1 in the plasma membrane can be verified at a much higher resolution by use of two-color immunofluorescence staining with super-resolution SIM (Fig. 2) [9]. In the present report, we describe the application of this method to reveal CAV1 localization in the plasma membrane with a significantly better spatial resolution than conventional microscopy.

2 Materials

All solutions are formulated with ultrapure water prepared with purified deionized water and analytical-grade reagents. Following preparation, all reagents are stored at room temperature, unless indicated otherwise. All waste disposal regulations must be diligently followed when disposing waste materials. In our laboratory, we do not add sodium azide to any reagent.

2.1 Immunofluorescence Staining Combined with Unroofing Method

1. Six-well plastic culture dish plates.
2. Select cell line(s) (NCI-H1975 lung adenocarcinoma cells in our experiments).
3. RPMI-1640 culture media containing 10% fetal bovine serum (FBS).
4. Micro-cover glass (coverslip, 18 × 18 mm).
5. Dulbecco's phosphate-buffered saline (PBS).
6. Nitrocellulose paper (cut at 18 × 18 mm for 6-well plate) (*see* **Note 1**).

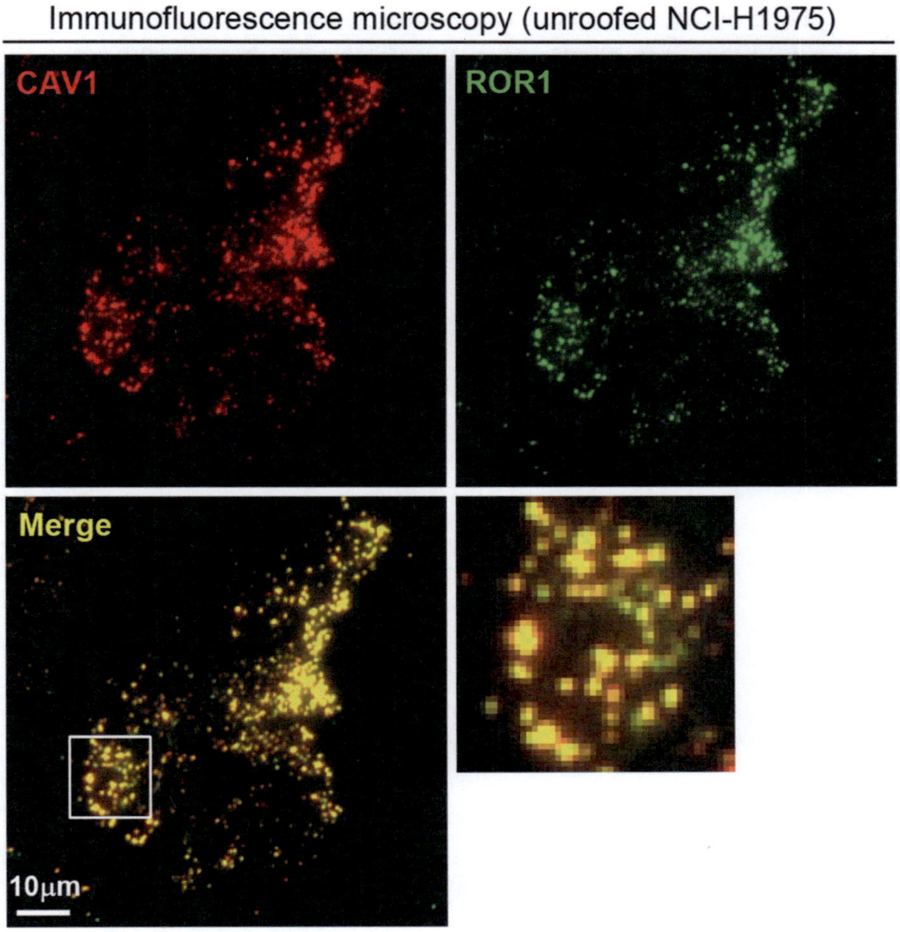

Fig. 1 Two-color immunofluorescence staining combined with the unroofing procedure revealed colocalization of punctate signals of CAV1 (red) with those of ROR1 (green) in NCI-H1975 lung adenocarcinoma cells

7. Fixation buffer:

   3.7% Formaldehyde freshly diluted in PBS: 1:10 stock solution (37%) (*see* **Note 2**)

8. Blocking buffer:

   1% Bovine serum albumin (BSA) diluted in PBS (*see* **Notes 3** and **4**).

9. Primary antibody: anti-CAV1 antibody, monoclonal mouse species, diluted to 10 μg/mL in PBS containing BSA solution. When performing double immunofluorescence staining, use an anti-CAV1 antibody, monoclonal mouse species, and anti-ROR1 antibody, polyclonal goat species, each diluted to 10 μg/mL in PBS containing BSA solution.

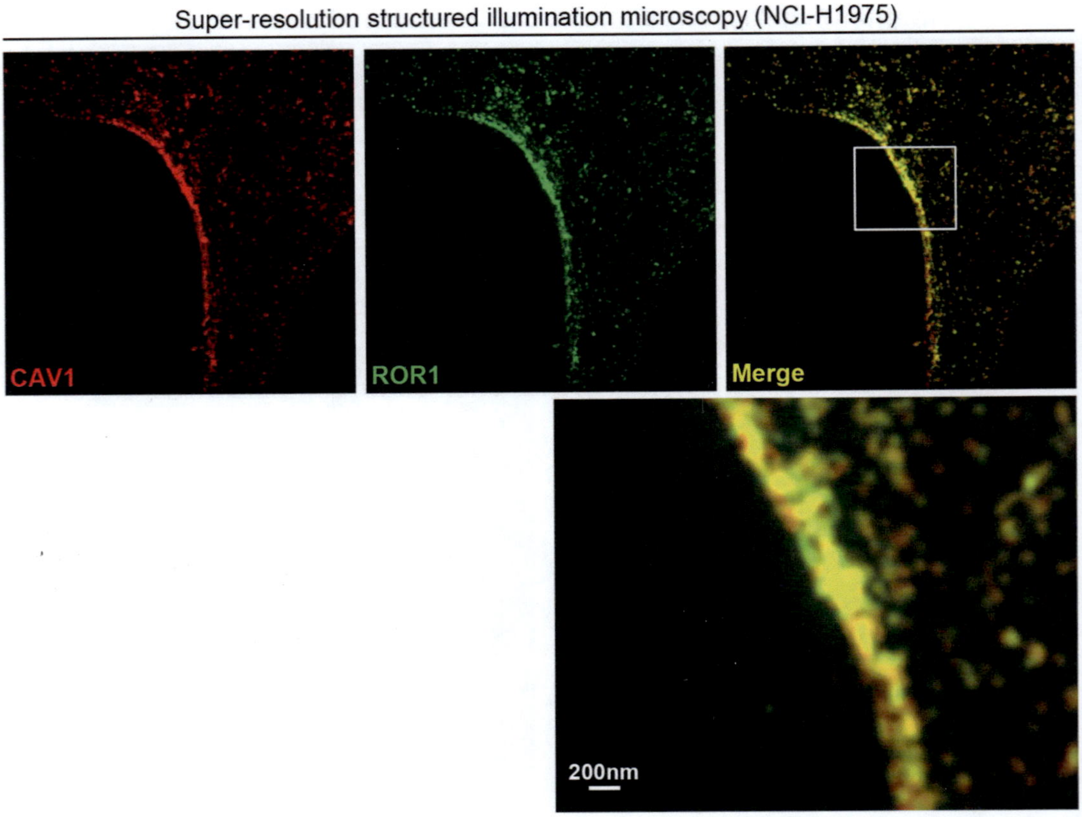

Fig. 2 Colocalization of CAV1 (red) with ROR1 (green) shown by two-color immunofluorescence staining using super-resolution SIM in NCI-H1975 lung adenocarcinoma cells

10. Secondary antibody: affinity-purified fluorescent antiglobulin conjugate reactive with species globulin for the first step, e.g., affinity-purified alexa-488-conjugated donkey antimouse IgG (H+L chains) diluted to 2.5 μg/mL in PBS containing BSA solution. When performing double immunofluorescence staining, use affinity-purified alexa-488-conjugated donkey antimouse IgG (H+L chains) and affinity-purified alexa-568-conjugated donkey antigoat IgG (H+L chains) antibodies, each diluted to 2.5 μg/mL in PBS containing BSA solution (*see* **Notes 5–8**).
11. Light protection box for suppression of fluorescence bleaching.
12. Adhesive glass slides (76 × 26 mm, thickness 0.9–1.2 mm).
13. Fluoromount (glycerol-mounting medium).
14. Fluorescence microscope equipped with appropriate filters (*see* **Note 9**).

2.2 Immunofluorescence Staining Using Super-Resolution SIM

1. Six-well plastic culture dish plates.
2. Select cell line(s) (NCI-H1975 lung adenocarcinoma cells in our experiments).
3. RPMI-1640 culture media containing 10% fetal bovine serum (FBS).
4. Marienfeld cover glass (Carl Zeiss high-performance ISO8255 compliant coverslip, 18x18 mm (length x width), No. 1.5 H, thickness 170 ± 5 μm).
5. Dulbecco's phosphate-buffered saline (PBS).
6. Fixation buffer:
   3.7% Formaldehyde freshly diluted in PBS: 1:10 stock solution (37%).
7. Permeabilization buffer:
   0.1% Triton X-100 freshly diluted in PBS: 1:10 stock solution (10%) (*see* **Note 10**).
8. Blocking buffer:
   1% Bovine serum albumin (BSA) diluted in PBS (*see* **Notes 3** and **4**).
9. Primary antibody: anti-CAV1 antibody, monoclonal mouse species, diluted to 10 μg/mL in PBS containing BSA solution. When performing double immunofluorescence staining, use an anti-CAV1 antibody, monoclonal mouse species, and anti-ROR1 antibody, polyclonal goat species, each diluted to 10 μg/mL in PBS containing BSA solution.
10. Secondary antibody: affinity-purified fluorescent antiglobulin conjugate reactive with the species globulin for the first step, e.g., affinity-purified alexa-488-conjugated donkey antimouse IgG (H+L chains) antibody diluted to 2.5 μg/mL in PBS containing BSA solution. When performing double immunofluorescence staining, use affinity-purified alexa-488-conjugated donkey antimouse IgG (H+L chains) and affinity-purified alexa-568-conjugated donkey antigoat IgG (H+L chains) antibodies, each diluted to 2.5 μg/mL in PBS containing BSA solution (*see* **Notes 5–8**).
11. Light-proof box for suppression of fluorescence bleaching.
12. Adhesive glass slides (76 × 26 mm, thickness 0.9–1.2 mm).
13. Vectashield antifade mounting medium (*see* **Note 11**).
14. Fluorescence microscope equipped with appropriate filters (*see* **Note 12**).

3 Methods

All procedures are performed at room temperature unless otherwise specified.

3.1 Immunofluorescence Staining Combined with Unroofing Method

1. NCI-H1975 lung adenocarcinoma cells maintained in RPMI-1640 with 10% fetal bovine serum.
2. 1.0×10^5 Cells plated onto coverslips in 6-well plates.
3. The next day, cells are unroofed mechanically using the method noted following (*see* **Note 13**) to obtain ventral plasma membrane and associated cytoskeleton components.
4. Wash cells 3 times with PBS, then overlay with prewet nitrocellulose paper (*see* **Note 1**).
5. Cells attached to the paper are removed by peeling off the paper (*see* **Note 14**).
6. Remaining membrane is immediately fixed with fixation buffer for 10 min.
7. Nonspecific binding is blocked by incubating coverslips for 30 min in blocking buffer.
8. Fixed cells are incubated with primary antibodies (anti-CAV1 (mouse) and anti-ROR1 (goat)) diluted in blocking buffer for 1 h.
9. Fixed cells are mildly washed with PBS 3 times.
10. Fixed cells are incubated with appropriate secondary antibodies conjugated with specific Alexa dye (Alexa-488 and Alexa-568) for 1 h using light protection box for suppression of fluorescence bleaching (*see* **Note 15**).
11. Fixed cells are mildly washed with PBS 3 times.
12. Mounting with Fluoromount (*see* **Note 16**).
13. Mounted slides are stored at 4 °C and protected from light.
14. Fluorescence acquisition is performed using AF6500 fluorescence microscope (Leica) and Zeiss LSM 880 confocal microscope (Carl Zeiss). Colocalization quantified using Image J software (*see* **Notes 17** and **18**).

3.2 Immunofluorescence Staining Using Super-Resolution SIM

1. NCI-H1975 lung adenocarcinoma cells maintained in RPMI-1640 with 10% fetal bovine serum.
2. 1.0×10^5 cells are plated onto coverslips in 6-well plates.
3. The next day, cells are washed 3 times with PBS and fixed with fixation buffer for 10 min (*see* **Note 19**).
4. Cells are permeabilized with permeabilization buffer, then washed 3 times with PBS.

5. Nonspecific binding is blocked by incubating coverslips for 30 min in blocking buffer.
6. Fixed cells are incubated with primary antibodies (anti-CAV1 (mouse) and anti-ROR1 (goat)) diluted in PBS containing 1.0% BSA for 1 h, then washed 3 times with PBS.
7. Fixed cells are incubated with appropriate secondary antibodies conjugated with specific Alexa dye (Alexa-488 and Alexa-568) for 1 h using light protection box for suppression of fluorescence bleaching.
8. Fixed cells are washed with PBS 3 times.
9. Mount coverslips with Vectashield (*see* **Notes 11** and **16**).
10. Coverslips can be sealed around the perimeter with nail polish.
11. Mounted slides are stored at 4 °C and protected from light.
12. Images are acquired with Zeiss LSM 880 confocal microscope using α Plan-Apochromat 100×/numerical aperture, 1.46 objective.
13. SIM images are collected using samples obtained with Zeiss ELYRA PS.1 system (Carl Zeiss Microscopy) with a 100× objective lens, numerical aperture 1.46, at room temperature. Three orientation angles and 5 phases of excitation grid are acquired for each *Z* plane, with *Z* spacing 167.2 nm between planes. SIM processing is performed using SIM module of ZEN 2 software (Carl Zeiss Microscopy) (*see* **Notes 20–22**).

4 Notes

1. Cut nitrocellulose paper into appropriate size, then immerse in PBS buffer using tweezers in advance [7].
2. Formaldehyde fixation is performed by crosslinking intracellular proteins.
3. The use of a competitor protein is important for these procedures to minimize nonspecific labeling because of sticking of the protein. BSA is a convenient and inexpensive purified protein available for this purpose, though other proteins, such as normal globulin of the same species as used in the immunoreaction step, or normal calf serum or plasma, are also useful for this purpose.
4. The blocking process is of particular importance for the unroofing method.
5. The use of an affinity-purified secondary antibody reaction reagent is important, since it dramatically reduces the background. Various companies offer such a reagent.

6. Alexa 488 dye is a bright green fluorescent dye with excitation ideally suited to the 488-nm laser line. For stable signal generation during imaging, Alexa Fluor 488 dye is pH insensitive over a wide molar range.
7. Alexa Fluor 568 dye is a bright orange fluorescent dye with excitation ideally suited to the 568-nm laser line with the Ar-Kr mixed-gas laser. For stable signal generation during imaging, Alexa Fluor 568 dye is pH insensitive over a wide molar range.
8. A probe with high fluorescence quantum yield and high photostability allows for detection of low-abundance biological structures with great sensitivity. Alexa Fluor 488 or 568 dye molecules can attach to proteins at high molar ratios without significant self-quenching, thus enabling brighter conjugates and more sensitive detection.
9. For our experiments, fluorescence is performed using an AF6500 fluorescence microscope (Leica) and Zeiss LSM880 confocal microscope (Carl Zeiss). Colocalization of CAV1 and ROR1 is quantified using ImageJ software. Pearson's correlation and a scatter diagram are determined by use of the colocalization plugin (e.g., CAV1 and ROR1) available for ImageJ.
10. Triton X-100 permeabilized membranes. Permeabilization is required when the antibody requires access to the inside of cells to detect the target antigen. Such antigens include intracellular proteins and cytoplasmic epitopes of transmembrane proteins. Solvents or detergents are typically used for permeabilization. Solvents can be used after fixation with a crosslinking agent such as formaldehyde, while detergents can either be used ionic (e.g., SDS), nonionic (e.g., TritonX-100, NP-40 or Octylglucoside), or zwitterionic (e.g., CHAPS). Detergent permeabilization can significantly improve antibody access to antigens in the cytoplasm or on the cytoplasmic face of the plasma membrane.
11. Vectashield antifade mounting medium is a stable formula useful for preserving fluorescence, and also prevents rapid photobleaching of fluorescent proteins and dyes [14]. To mount cells onto a slide, dispense Vectashield onto the specimen using a disposable pipette or pipette tip. Drops with a small volume of approximately 10 μL (with an 18 × 18 mm coverslip) are recommended. Coverslip the specimen and allow the Vectashield to become dispersed over the entire section. Tightly seal the bottle with the screw cap after use and store in an upright position. Vectashield does not solidify, but remains a liquid while on the slide and can be stored without sealing. If desired, coverslips can be sealed around the perimeter with nail polish or plastic sealant. Mounted slides should be stored at 4 °C and protected from light.

12. In our experiments, fluorescence is performed using a Zeiss LSM880 confocal microscope with the Zeiss ELYLA PS.1 system (Carl Zeiss).
13. Cell density should range from 80% to 90% when performing an unroofing procedure.
14. When performed correctly, phase-contrast microscopy will show that 60–70% of the cells have become successfully unroofed, leaving the ventral plasma membrane and some cytoplasmic structures.
15. By immunolabeling the ventral plasma membrane preparation, the distributions of CAV1 and ROR1 can be clearly observed in a single focal plane.
16. Be careful not to allow bubbles to form between the coverslip and cover glass.
17. With such specimens, labeling for CAV1 is observed as dots in a single focal plane, which most likely represents adherence to the basal plasma membrane. Each dot appears to be independent and typically has a clear focus.
18. We have used an unroofing procedure with immunofluorescence microscopy employed to analyze basal plasma membrane localization of CAV1 and ROR1, which clearly showed their colocalization in a single plane of the adhered membrane (Fig. 1). Our findings demonstrated that CAV1 and ROR1 colocalize in caveolae structures in lung adenocarcinoma cells [9].
19. Cell density should be 70–80% when performing the fixing procedure.
20. Structured illumination is used to enhance spatial resolution. It involves illuminating the sample with patterned light and then using software to analyze information in Moiré fringes outside the normal range of observation. Reconstruction software deciphers the images at about twofold higher resolution than the diffraction limit, or approximately 100 nm. SIM has advantages over other super-resolution microscopy methods in that it can be used for imaging thicker sections, as well as 3D imaging and live-cell imaging. Image quality increases with the use of bright and photostable dyes. Fluorescent proteins are commonly used for SIM investigations of live cells, in addition to multiplexing with organic dyes and Qdot probes. Applications include investigation of the membrane nanostructure, protein aggregation, nuclear machinery, molecular architecture of the cell–cell interface, and synaptic transmission [12, 13].

21. The detailed procedures and manual for the Carl Zeiss LSM880 ELYRA PS.1 system can be obtained from the Carl Zeiss homepage (https://www.zeiss.com/microscopy/int/home.html).
22. In our study, colocalization of CAV1 and ROR1 was verified at a much higher resolution by two-color immunofluorescence staining using super-resolution SIM [9] (Fig. 2).

Acknowledgments

This work was supported in part by Grants-in-Aid for Scientific Research (A) and (C) from the Japan Society for the Promotion of Science (JSPS), and Project for Cancer Research and Therapeutic Evolution (P-CREATE) program of the Japan Agency for Medical Research and Development (AMED), as well as a Grant-in-Aid for Scientific Research on Innovative Areas from the Ministry of Education, Culture, Sports, Science and Technology (MEXT) of Japan.

References

1. Parton RG, del Pozo MA (2013) Caveolae as plasma membrane sensors, protectors and organizers. Nat Rev Mol Cell Biol 14:98–112
2. Parton RG, Simons K (2007) The multiple faces of caveolae. Nat Rev Mol Cell Biol 8:185–194
3. Hansen CG, Nichols BJ (2010) Exploring the caves: cavins, caveolins and caveolae. Trends Cell Biol 20:177–186
4. Lamaze C, Tardif N, Dewulf M, Vassilopoulos S, Blouin CM (2017) The caveolae dress code: structure and signaling. Curr Opin Cell Biol 47:117–125
5. Fujimoto T (1993) Calcium pump of the plasma membrane is localized in caveolae. J Cell Biol 120:1147–1157
6. Fujimoto T, Fujimoto K (1997) Metal sandwich method to quick-freeze monolayer cultured cells for freeze-fracture. J Histochem Cytochem 45:595–598
7. Fujimoto T, Lee K, Miwa S, Ogawa K (1991) Immunocytochemical localization of fodrin and ankyrin in bovine chromaffin cells in vitro. J Histochem Cytochem 39:1485–1493
8. Fujimoto T, Ogawa K (1989) Immunoelectron microscopy of fodrin in the rat uriniferous and collecting tubular epithelium. J Histochem Cytochem 37:1345–1352
9. Yamaguchi T, Lu C, Ida L, Yanagisawa K, Usukura J, Cheng J, Hotta N, Shimada Y, Isomura H, Suzuki M, Fujimoto T, Takahashi T (2016) ROR1 sustains caveolae and survival signalling as a scaffold of cavin-1 and caveolin-1. Nat Commun 7:10060
10. Yamaguchi T, Yanagisawa K, Sugiyama R, Hosono Y, Shimada Y, Arima C, Kato S, Tomida S, Suzuki M, Osada H, Takahashi T (2012) NKX2-1/TITF1/TTF-1-induced ROR1 is required to sustain EGFR survival signaling in lung adenocarcinoma. Cancer Cell 21:348–361
11. Schermelleh L, Heintzmann R, Leonhardt H (2010) A guide to super-resolution fluorescence microscopy. J Cell Biol 190:165–175
12. Leung BO, Chou KC (2011) Review of super-resolution fluorescence microscopy for biology. Appl Spectrosc 65:967–980
13. Gustafsson MG, Shao L, Carlton PM, Wang CJ, Golubovskaya IN, Cande WZ, Agard DA, Sedat JW (2008) Three-dimensional resolution doubling in wide-field fluorescence microscopy by structured illumination. Biophys J 94:4957–4970
14. Krenik KD, Kephart GM, Offord KP, Dunnette SL, Gleich GJ (1989) Comparison of antifading agents used in immunofluorescence. J Immunol Methods 117:91–97

Chapter 5

Quantitative Image Analysis of the Spatial Organization and Mobility of Caveolin Aggregates at the Plasma Membrane

Takashi Hirama and Raibatak Das

Abstract

Caveolins are integral membrane proteins that are the principal structural component of caveolae. Newly synthesized caveolin self-associates into oligomers that further assemble into higher-order structures. Imaging fluorescently labeled caveolin at the plasma membrane with total internal reflection fluorescence (TIRF) microscopy reveals a spatially heterogeneous distribution with aggregates of various sizes. In this chapter, we present a set of image-processing tools to quantify the spatial organization and mobility of caveolin aggregates seen in TIRF images. We apply a spot detection algorithm to identify punctate features on multiple length scales, and computationally estimate the area and integrated fluorescence signal of each detected feature. We then partition the original image into two disjoint sets: one containing pixels within punctae, and the other containing pixels on the rest of the plasma membrane. From these partitions, we estimate the relative fraction of caveolin that is punctate versus diffuse. Finally, we analyze the mobility of caveolin aggregates by tracking them and classify individual trajectories as diffusive or subdiffusive using a moment scaling spectrum analysis. Together, these analyses capture multiple facets of caveolin organization and dynamics. To demonstrate their utility, we quantify the distribution of fluorescent Caveolin 1 stably transfected in HeLa cells. We analyze cells at baseline and after being exposed to the anesthetic Dibucaine that is known to scramble membrane phospholipids. Our analysis shows how this perturbation dramatically alters caveolin aggregation and mobility.

Key words Caveolin, TIRF, Spot detection, Spatial heterogeneity, Moment scaling spectrum

1 Introduction

Caveolae are flask-shaped nanodomains of the plasma membrane enriched in cholesterol, sphingolipids, and anionic phospholipids, including phosphatidylserine (PtdSer) or phosphatidylinositol 4,5-bisphosphate (PtdIns(4,5)P_2) [1–3]. A variety of physiological roles for caveolae have been described, including lipid transport, endocytic traffic, and membrane stress regulation. Caveolin is an integral membrane protein and a fundamental component in caveolae formation. Caveolin 1 (Cav1) is the most widely studied isoform of this protein. By imaging fluorescently tagged Cav 1, caveolae are shown to be punctate structures on the plasma membrane (Fig. 1a).

Cedric M. Blouin (ed.), *Caveolae: Methods and Protocols*, Methods in Molecular Biology, vol. 2169,
https://doi.org/10.1007/978-1-0716-0732-9_5,

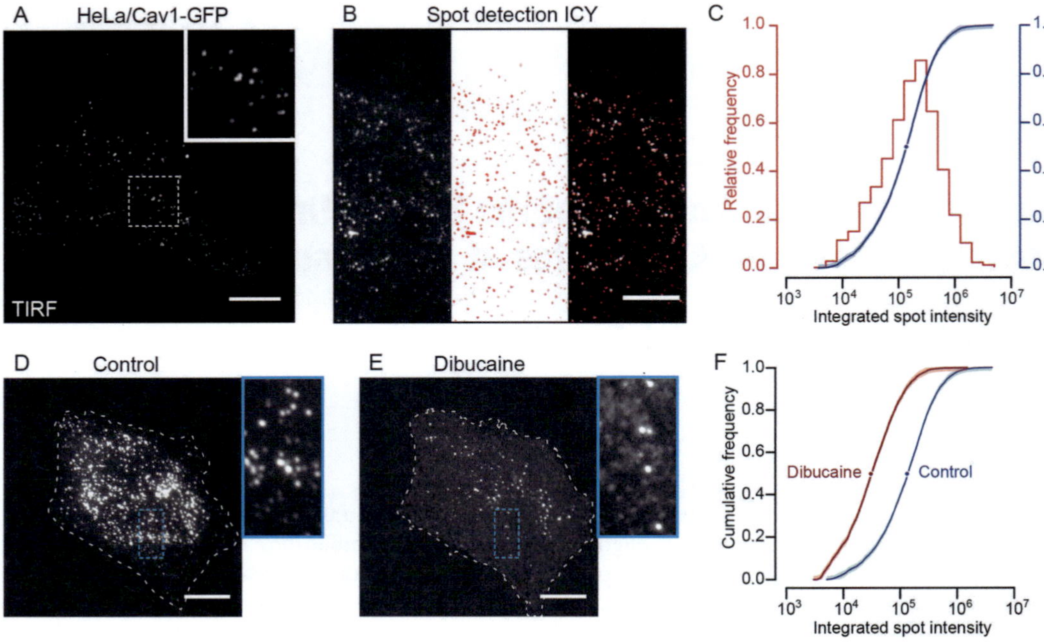

Fig. 1 (**a**) Representative TIRF image of HeLa/Cav1-GFP. Insets: magnification of the region of interest (dashed square). Scale bar = 10 μm unless stated. (**b**) Detection of Cav1-GFP punctum in TIRF images using Spot Detector function in Icy. The raw TIRF image (left), the detected Cav1 by spot detector (center), and the overlay (right). (**c**) Relative and the empirical cumulative frequency distribution (ECDF, or CDF) of the integrated spot intensities in detected features of Cav1-GFP in HeLa cells ($n = 36$ cells). The circle in the middle is the median value and the shaded region along CDF is 95% confidence band. (**d**) and (**e**) Representative TIRF images of HeLa/Cav1-GFP before (**d**) and after (**e**) dibucaine treatment. The dashed line in white indicates the cell margin. Insets, magnifications of the indicated area by blue square. (**f**) The CDF of integrated spot intensities in HeLa/Cav1-GFP before (blue) and after (red) dibucaine treatment ($n = 18$ cells). (Adapted from [4] with permission from *The Journal of Biological Chemistry*)

As the majority of caveolin resides on the plasma membrane [5], total internal reflection fluorescence (TIRF) microscopy is ideal to visualize its spatial distribution. The TIRF evanescent excitation field penetrates the specimen about 100 nm from the coverslip surface and thus allows Cav1 embedded in caveolae to be clearly imaged with minimal intracellular background signal.

We recently demonstrated that anionic phospholipids, the majority of which are asymmetrically localized in the cytosolic leaflet of the plasma membrane [6, 7], dictate the assembly and stability of caveolae [4]. This chapter describes quantitative image analysis methods we develop to:

1. Quantify the size and intensity of punctate Cav1 structures seen in TIRF images,
2. Measure the relative amounts of punctate versus diffuse fluorescent signal on the plasma membrane, and,
3. Analyze the mobility of Cav1 aggregates on the membrane.

We apply these methods to cells stably expressing Cav1-GFP and measure the effect of treatment with the anesthetic Dibucaine that is known to scramble phospholipids in the plasma membrane. Dibucaine-induced scrambling alters the phospholipid asymmetry. Our analysis shows that this results in a significant reduction of Cav1-GFP punctae, leads to the disassembly of GFP signal from Cav1 punctae in the plane of the plasma membrane, and increases the mobility of Cav1 aggregates, and reduces their apparent confinement [4]. Each analysis highlights unique changes in Cav1 features, morphology, and dynamics upon perturbation.

2 Materials

2.1 Cell Culture

1. HeLa cell line (the American Type Culture Collection).
2. DMEM/F12 medium.
3. Fetal bovine serum (FBS) (5% final).
4. Cav1-mEGFP (Addgene).
5. pMSCV-puro vector (Clontech),
6. FuGene6 (Promega).
7. Puromycin (1 μg/ml final).
8. Influx BRV (BD Biosciences).
9. Glass coverslips with thickness No. 1.5H.Forward primer for Cav1-mEGFP 5′-GCGAGATCTATGTCTGGGGGCAAATACGTAG-3′.
10. Reverse primer for Cav1-mEGFP 5′-GCGGAATTCTTACTT GTACAGCTCGTCCATG-3′.

2.2 TIRF Microscope

1. Olympus IX81 inverted microscope.
2. Hamamatsu C9100–13 EM-CCD.
3. 150×/1.45 oil immersion objectives
4. 491- and 561-nm laser excitation lines
5. Volocity 4.3.2 software (Perkin Elmer).
6. A magnetic Chamlide chamber (Chamlide CMB, Live Cell Instrument).

2.3 Reagent

1. Dibucaine (1 mM final).
2. Medium A:
   140 mM NaCl, 5 mM KCl, 1 mM $MgCl_2$, 20 mM HEPES, 100 μM EGTA, 25 mM glucose, pH 7.4, adjusted with KOH.

2.4 Soft- and Hardware for Post-acquisition Analysis (See Note 1)

1. ImageJ (National Institutes of Health, Bethesda, MD).
2. ICY (Institut Pasteur, Paris, France).
3. MATLAB—including Image Processing toolbox (MathWorks, Natick, MA).
4. Single-particle-tracking code for MATLAB.
5. Spot intensity analysis code in MATLAB, freely available for download at https://gitlab.com/dododas/caveolae

3 Methods

3.1 Cell Line Stably Expressing Cav1-GFP

1. PCR is performed, based on Cav1-mEGFP with primers described in Subheading 2.1 by starting at 95 °C for 30 s, followed by 40 cycles of 95 °C for 8 s, 81 °C for 25 s, and 72 °C for 20 s [8].
2. Subcloning is then done using BglII and EcoRI restriction sites.
3. The PCR product is inserted into pMSCV-puro vector to construct pMSCV-Cav1-GFP plasmid.
4. HeLa cells are incubated in DMEM/F12 with 5% FBS at 37 °C under 5% CO_2 and routinely grown in T-25 tissue culture flasks.
5. The cells are transfected with pMSCV-cav1-GFP by using FuGene6, maintained for 3 weeks in DMEM/F12 containing 1 μg/ml puromycin.
6. The cells are then sorted with Influx BRV, followed by incubation in DMEM/F12 with 5% FBS.

3.2 Image Acquisition Through TIRF Microscope

1. HeLa cells expressing Cav1-GFP (HeLa/Cav1-GFP) are deposited on glass coverslips into 12-well plates 24 h before microscopy assay.
2. After placing the coverslip into a magnetic Chamlide chamber, HeLa/Cav1-GFP are imaged with TIRF microscope.
3. For the acquisition of images or movies (500 frames over 20 s), prewarmed medium A is replaced and the chamber is placed on a microscope stage maintained at 37 °C.

3.3 Detection of Cav1-GFP Punctum as Spot Areas and Intensities

Cav1-GFP punctae are identified using the automated spot detection plugin in ICY. The distribution of spot areas and mean pixel intensities are computationally calculated using a custom MATLAB script spot_analysis_script.m (available in the online repository, also *see* **Notes 2** and **3**). The script requires three source images: the original Cav-1 GFP image, a binary membrane mask, and the spots mask generated by ICY. It applies morphological analysis tools from the MATLAB Image Processing toolbox to compute the area, the mean intensity, and their product for each detected spot. The product is defined as an integrated spot intensity. It is essentially the total fluorescence signal within each spot. The script computes these metrics for each spot, plots their distribution, and saves the results into a csv file for further analysis.

To visualize the integrated intensity distribution, we typically plot its empirical cumulative distribution function (ECDF). Changes in the distribution, for example, in response to pharmacological treatment, manifest as shifts in the ECDF plot (*see* **Notes 4** and **5**).

1. Fluorescence signal for stably expressed Cav1-GFP in HeLa cells imaged using a TIRF microscope is shown in Fig. 1a. Punctate features of varying sizes and intensities are visually apparent. These features are extracted with an automated spot detection program in the free software ICY. The spot detection algorithm uses a multiscale product of wavelet transforms of the original image to extract features over multiple length scales [9] (Fig. 1b).
2. The pixel area, mean pixel intensity, and area the integrated spot intensity for each detected spot is computed using the included MATLAB script. The frequency distribution of integrated spot intensity is shown in the histogram in red and the cumulative frequency distribution in blue (Fig. 1c).
3. When membrane lipids are scrambled by dibucaine for 10 min at 37 °C in medium A (Fig. 1d), the abundance of Cav1-GFP punctae decreased as seen through TIRF imaging (Fig. 1e).
4. Accordingly, the integrated spot intensity distribution shifts to the left upon dibucaine treatment (Fig. 1f).

3.4 Quantifying Aggregated vs Diffuse Caveolin

We analyze the relative distribution of GFP fluorescence between punctate structures versus diffuse on the membrane. This analysis measures the dispersion of Cav1-GFP in the plane of the membrane. The included MATLAB script uses the input masks to split the original Cav1-GFP image into two nonoverlapping components—an image containing just the punctate signal, and an image containing the remaining diffuse signal at the plasma membrane. From these partitioned images, it computes the ratio of total signal that is diffuse versus punctate. A higher ratio means the Cav1-GFP is relatively more dispersed. The script saves the partitioned images as tif files, and the intensity analysis results in a csv file for further analysis.

1. The original TIRF image (Fig. 2a) is converted to two tif files: one for a membrane mask and the other for identifying Cav1-GFP punctum. (8-bit, contrast-enhanced version of the original 16-bit images are shown here for clarity. The original images are available in the online repository).

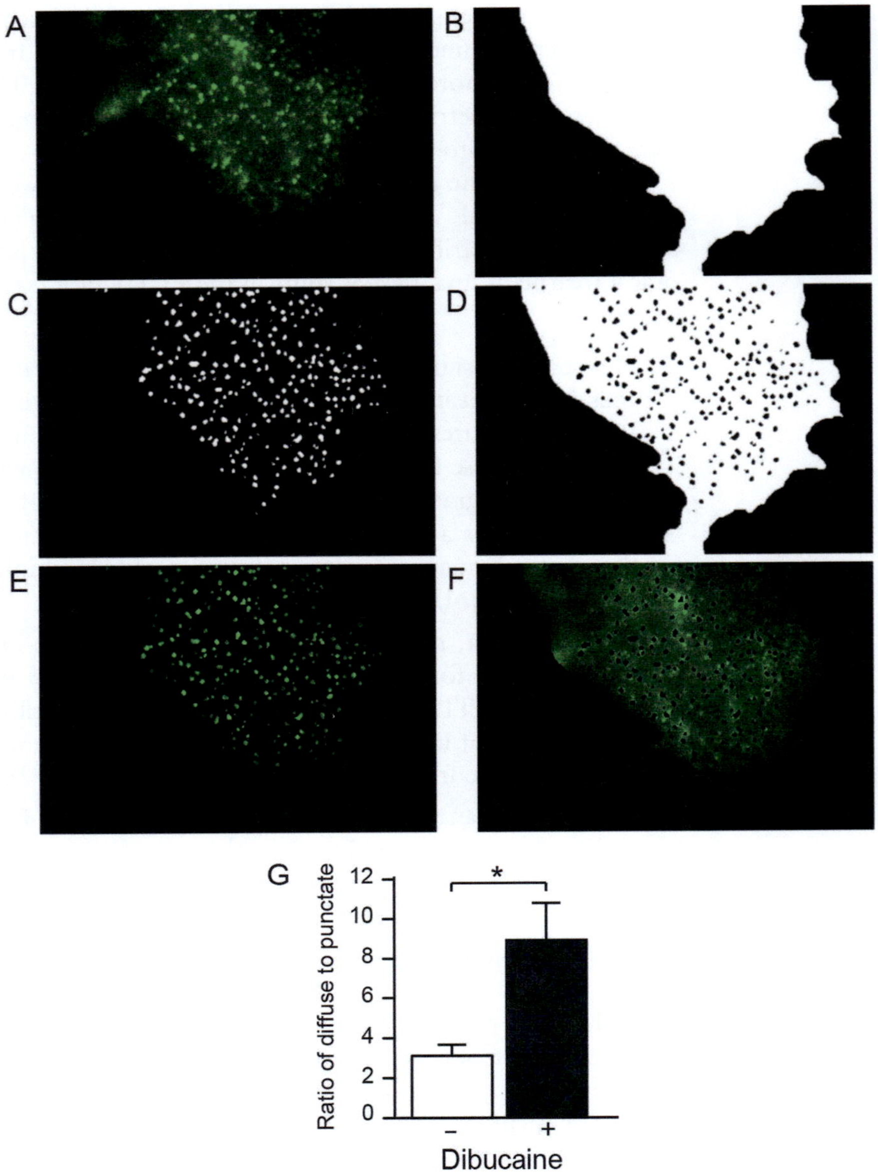

Fig. 2 (**a**) TIRF image of HeLa/Cav1-GFP that is used to generate membrane and spot masks. (8-bit, contrast-enhanced version of the original 16-bit images is shown here for clarity. The original images are available in the online repository). (**b**) Membrane mask image generated by the original image using imageJ. (**c**) Spot mask image generated by spot detection ICY to identify Cav1-GFP punctum, and combined with the membrane mask to retain only spots that are on the cell of interest. (**d**) Combined mask images from the membrane mask (**b**) and the inverse of (**c**) to identify the diffuse GFP signal. (**e**) Partitioned image of Cav1-GFP in punctate form. (8-bit, contrast-enhanced version of the original 16-bit images is shown here for clarity. The original images are available in the online repository). (**f**) Partitioned image of Cav1-GFP in diffuse (8-bit, contrast-enhanced version of the original 16-bit images is shown here for clarity. The original images are available in the online repository). (**g**) Ratio of diffuse to punctate fluorescence (mean $\pm$ S.E.M. $n = 16$ cells). (Adapted from [4] with permission from *The Journal of Biological Chemistry*)

2. The membrane mask (binary image) is constructed manually by maximizing the GFP signal in imageJ to view the cell margin filling any holes (Fig. 2b).
3. Cav1-GFP punctum is identified by the same means as previously described in 3.3 (spot detection algorithm in ICY), and combined with the membrane mask (Fig. 2b) to generate a mask for detected spots at the membrane (Fig. 2c).
4. The membrane mask (Fig. 2b) is then combined with the inverse of the spot detection output (Fig. 2c) to generate a mask for diffuse Cav1-GFP at the membrane (Fig. 2d).
5. Applying the two masks to the original image (Fig. 2a) splits the original image of Cav1-GFP into either punctate (Fig. 2e) and diffuse partitions (Fig. 2f) (8-bit, contrast-enhanced version of the original 16-bit images are shown here for clarity. The original images are available in the online repository).
6. The total signal in each partition is calculated by summing the individual pixel intensities.
7. The ratio of diffuse to punctate fluorescence in untreated HeLa/Cav1-GFP cells (3.1) and cells treated with dibucaine (8.9) is shown in Fig. 2g. The visually diffuse GFP fluorescence in dibucaine-treated cells is reflected in a significantly higher diffuse-to-punctate ratio.

3.5 Diffusional Analysis of Cav1-GFP Features in HeLa/Cav1-GFP

Caveolae are generally immobile in unstimulated cells, and their mobility and trajectory are quantified by single-particle tracking (SPT) analysis. The type of motion (subdiffusive motion or free motion) and the diffusion coefficient ($\mu m^2/s$) are determined using a moment-scaling spectrum via a MATLAB algorithm. The trajectory of Cav1-GFP punctum in untreated cells is compared with that of dibucaine treatment (*see* **Notes 4** and **5**).

1. The motion of Cav1-GFP punctum in the plasma membrane is acquired by time-lapse capture (25 acquisitions/s for a total of 20 s) on TIRF microscope.
2. The trajectory of individual Cav1-GFP particles is analyzed by single-particle tracking (SPT).
3. A region of interest is cropped in a set of images (500 frames of tif pictures) by ImageJ (Fig. 3a).
4. Tracking is carried out in MATLAB; discerning local intensity maxima and fitting two-dimensional Gaussian kernels (refer to the code for the single-particle tracking [10]).
5. The particles identified through the time-lapse TIRF images are linked between consecutive pictures, and the complete trajectories are generated from the track segments by closing gaps and capturing merging and splitting events (Fig. 3b, d).

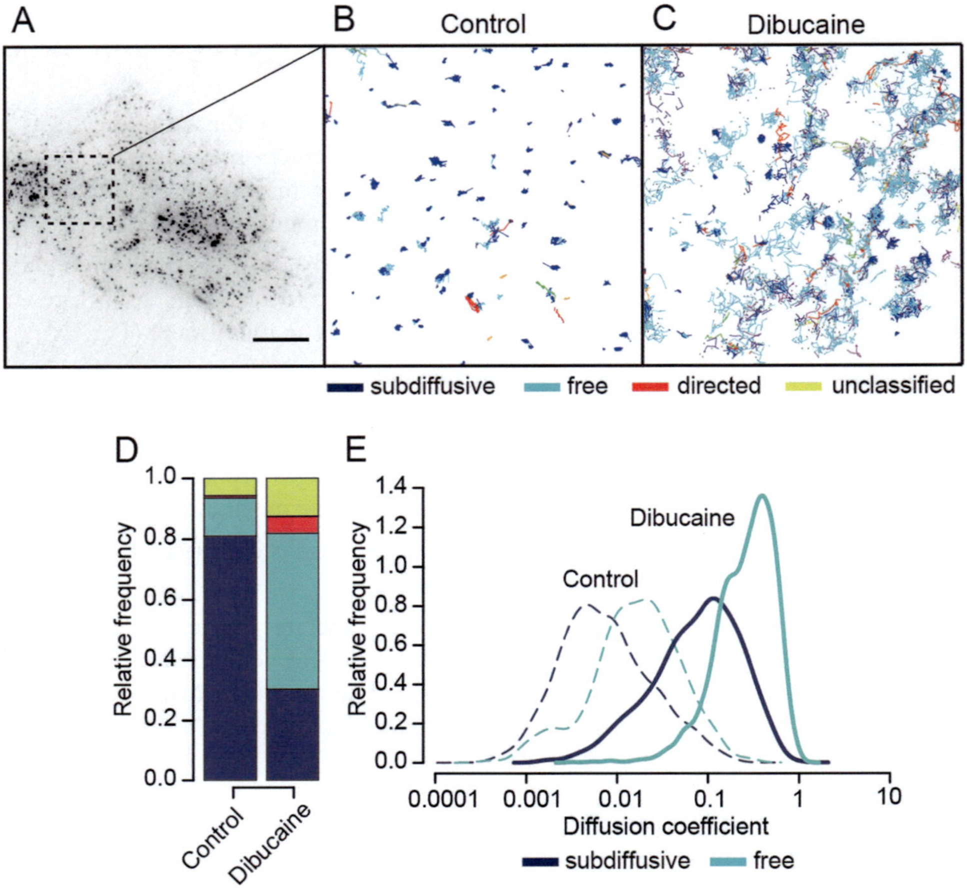

Fig. 3 (**a**) Inverted TIRF image of HeLa/Cav1-GFP recorded at a rate of 500 frames over 20 s. (**b**) Representative image of the tracking for caveolin with the tracks color-coded as follows: subdiffusive-Brownian trajectories in blue; pure-Brownian trajectories in cyan; linear trajectories in red; and unclassified trajectories in yellow. (**c**) Representative trajectories of caveolin features in cells treated with dibucaine. (**d**) Classification of the trajectories of caveolin features in cells untreated (control) and treated with dibucaine. (**e**) The relative frequency of the diffusion coefficients for the "subdiffusive" movement in blue, and "free" motion in cyan of caveolin. The dashed line indicates the untreated cells and the solid line indicates that of treatment with dibucaine. (Adapted from [4] with permission from *The Journal of Biological Chemistry*)

6. SPT analysis demonstrates that 81.4% of the Cav1-GFP punctum displayed subdiffusive motion, while free motion comprised 12.1% of the caveolae (Fig. 3d) consistent with previous reports [5, 11].
7. A significant drop in the fraction of subdiffusive motion is seen at 30.8% after dibucaine treatment, while Cav1-GFP punctum appears to be free motion increased to 50.4% with dibucaine treatment (Fig. 3c).
8. Motion types and diffusion coefficients are also derived with MATLAB using a moment scaling spectrum (MSS) analysis [12].

9. The confinement dimension for diffusion coefficients is calculated through eigenvalue decomposition of the variance-covariance matrix of particle positions [10].
10. The subdiffusive Cav1-GFP punctum has reduced diffusion coefficients compared to the free Cav1-GFP punctum with a median diffusion coefficient of 6.7×10^{-3} $\mu m^2/s$ compared to 1.7×10^{-2} $\mu m^2/s$ (Fig. 3e).
11. Dibucaine treatment changes the proportion of diffusion coefficients in Cav1-GFP punctum to 87.0×10^{-3} $\mu m^2/s$ of subdiffusive motion and 29.0×10^{-2} $\mu m^2/s$ of free motion shown in Fig. 3e.

4 Notes

1. The computational tools to quantify the spatial distribution of the membrane protein Caveolin through fluorescence microscopy images are obtainable and easy to handle.
2. The online repository uses the git version control system and is hosted on gitlab.com. The landing page is a README file with step-by-step instructions on how to download and use the included MATLAB script.
3. A python version is under development. Please contact the authors for more details.
4. The combination of TIRF microscope and SPT analysis is a powerful tool to quantify the lateral motion of mobile particles on the plane of the plasma membrane.
5. The utility of the methods described here is not restricted to Cav1-GFP. These analyses can be used to quantify the spatial distribution and mobility of other heterogeneously distributed membrane proteins.

Acknowledgments

We thank Gregory D. Fairn, Keenan Research Centre for Biomedical Science, St. Michael's Hospital, Toronto, Canada, and Sergio Grinstein, Program in Cell Biology, The Hospital for Sick Children, Toronto, Canada, for their support and for guiding the research presented here. This work is supported in part by Kurozumi Medical Foundation and Tokyo-Hokenkai Byotai-Seiri Laboratory.

References

1. Parton RG, Del Pozo MA (2013) Caveolae as plasma membrane sensors, protectors and organizers. Nat Rev Mol Cell Biol 14:98–112
2. Fairn GD, Schieber NL, Ariotti N, Murphy S, Kuerschner L, Webb RI, Grinstein S, Parton RG (2011) High-resolution mapping reveals topologically distinct cellular pools of phosphatidylserine. J Cell Biol 194:257–275
3. Fujita A, Cheng J, Tauchi-Sato K, Takenawa T, Fujimoto T (2009) A distinct pool of phosphatidylinositol 4,5-bisphosphate in caveolae revealed by a nanoscale labeling technique. Proc Natl Acad Sci 106:9256–9261
4. Hirama T, Das R, Yang Y, Ferguson C, Won A, Yip CM, Kay JG, Grinstein S, Parton RG, Fairn GD (2017) Phosphatidylserine dictates the assembly and dynamics of caveolae in the plasma membrane. J Biol Chem 292:14292. https://doi.org/10.1074/jbc.M117.791400
5. Pelkmans L, Zerial M (2005) Kinase-regulated quantal assemblies and kiss-and-run recycling of caveolae. Nature 436:128–133
6. Hirama T, Lu SM, Kay JG, Maekawa M, Kozlov MM, Grinstein S, Fairn GD (2017) Membrane curvature induced by proximity of anionic phospholipids can initiate endocytosis. Nat Commun 8:1393
7. Hirama T, Fairn GD (2018) Induction of spontaneous curvature and endocytosis: unwanted consequences of cholesterol extraction using methyl-β-Cyclodextrin. Commun Integr Biol 11:1–4. https://doi.org/10.1080/19420889.2018.1444306
8. Hirama T (2016) A real-time PCR-based diagnostic test for organisms in respiratory tract infection. INTECH. https://doi.org/10.5772/65740
9. Olivo-Marin J-C (2002) Extraction of spots in biological images using multiscale products. Pattern Recogn 35:1989–1996
10. Jaqaman K, Loerke D, Mettlen M, Kuwata H, Grinstein S, Schmid SL, Danuser G (2008) Robust single-particle tracking in live-cell time-lapse sequences. Nat Methods 5:695–702
11. Tagawa A, Mezzacasa A, Hayer A, Longatti A, Pelkmans L, Helenius A (2005) Assembly and trafficking of caveolar domains in the cell: caveolae as stable, cargo-triggered, vesicular transporters. J Cell Biol 170:769–779
12. Ferrari R, Manfroi AJ, Young WR (2001) Strongly and weakly self-similar diffusion. Phys D Nonlinear Phenom 154:111–137

Chapter 6

Spatiotemporal Analysis of Caveolae Dynamics Using Total Internal Reflection Fluorescence Microscopy

Yosuke Senju and Shiro Suetsugu

Abstract

Total internal reflection fluorescence microscopy enables to analyze the localizations and dynamics of cellular events that occur at or near the plasma membrane. Total internal reflection fluorescence microscopy exclusively illuminates molecules in the close vicinity of the glass surface, thereby reducing background fluorescence and enabling observation of the plasma membrane in the glass-attached cells with a high signal-to-noise ratio. Here, we describe the application of total internal reflection fluorescence microscopy to analyze the dynamics of caveolae, which play essential physiological functions, including membrane tension buffering, endocytosis, and signaling at the plasma membrane.

Key words Total internal reflection fluorescence microscopy, Kymograph, Endocytosis, Caveolae, Lateral diffusion

1 Introduction

Conventional wide-field and confocal microscopes have been widely used to image subcellular structures; however, they also give background fluorescence from outside of the focal plane. Total internal reflection fluorescence (TIRF) microscopy selectively excites fluorescent molecules in the close vicinity of the glass surface, where the cell attaches through the extracellular matrix. By eliminating fluorescence from intracellular regions with reducing background fluorescence, TIRF microscopy brings numerous benefits to the spatiotemporal analysis of molecular dynamics at or near the basement plasma membrane, even at the single-molecule level [1].

Total internal reflection is an optical phenomenon occurring at the interface between two mediums with different reflective indices (Fig. 1). When light propagates from a medium with a higher refractive index to a medium with a lower refractive index, it is not reflected if the incident angle θ is greater than the critical angle θ_c, and undergoes total reflection by generating evanescent waves

Cedric M. Blouin (ed.), *Caveolae: Methods and Protocols*, Methods in Molecular Biology, vol. 2169,
https://doi.org/10.1007/978-1-0716-0732-9_6, © Springer Science+Business Media, LLC, part of Springer Nature 2020

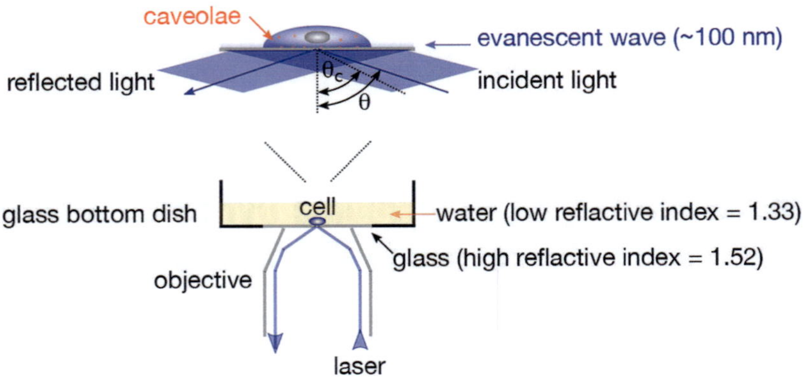

Fig. 1 Schematic representation of total internal reflection fluorescence (TIRF) microscopy. Transfected cells are cultured on a glass-bottom dish. The pH and temperature of cell-culture medium are controlled. When the excitation laser passes through the cover glass beyond the critical angle θ_C, the light undergoes total internal reflection and generates an evanescent wave that penetrates approximately 100–200 nm into the medium. Consequently, the evanescent wave selectively excites caveolae at or near the basement plasma membrane with reducing background fluorescence. The focus is adjusted during TIRF observation using the Nikon Perfect Focus System

toward the medium with the lower refractive index. The intensity of the evanescent wave decays exponentially as a function of the distance from the interface. In the case of TIRF microscopy, total reflection occurs at the interface between the cover glass (reflective index, $n = 1.52$) and cell-culture medium (reflective index, $n = 1.33$). Hence, a high numerical aperture (NA) objective is required to generate total reflection. The penetration depth of this evanescent field is typically 100–200 nm. Consequently, the evanescent wave selectively excites fluorescence molecules at or near the basement plasma membrane in cells. This spatially confined region of excitation by evanescent wave significantly increases the signal-to-noise (S/N) ratio of the fluorescence images. This low-energy evanescent wave is also beneficial for live-cell imaging for a longer period of time by reducing phototoxicity to the cells.

Because of these optical features of TIRF microscopy, it is suitable for studying dynamic cellular processes that occur at or near the plasma membrane (e.g., caveolae biogenesis, clathrin-dependent and -independent endocytosis, ligand-receptor interaction, and focal adhesion turnover) [2–7]. Here, we show the TIRF microscopy set-up (Fig. 1) and its application to study the spatiotemporal dynamics of caveolae, by generating a kymograph from time-lapse TIRF images and estimating the lifetime of caveolae markers at the plasma membrane (Fig. 2) [8, 9]. TIRF microscopy will continue to provide new insights in identifying proteins involved in caveolae and determining the order in which caveolae components are recruited to the plasma membrane.

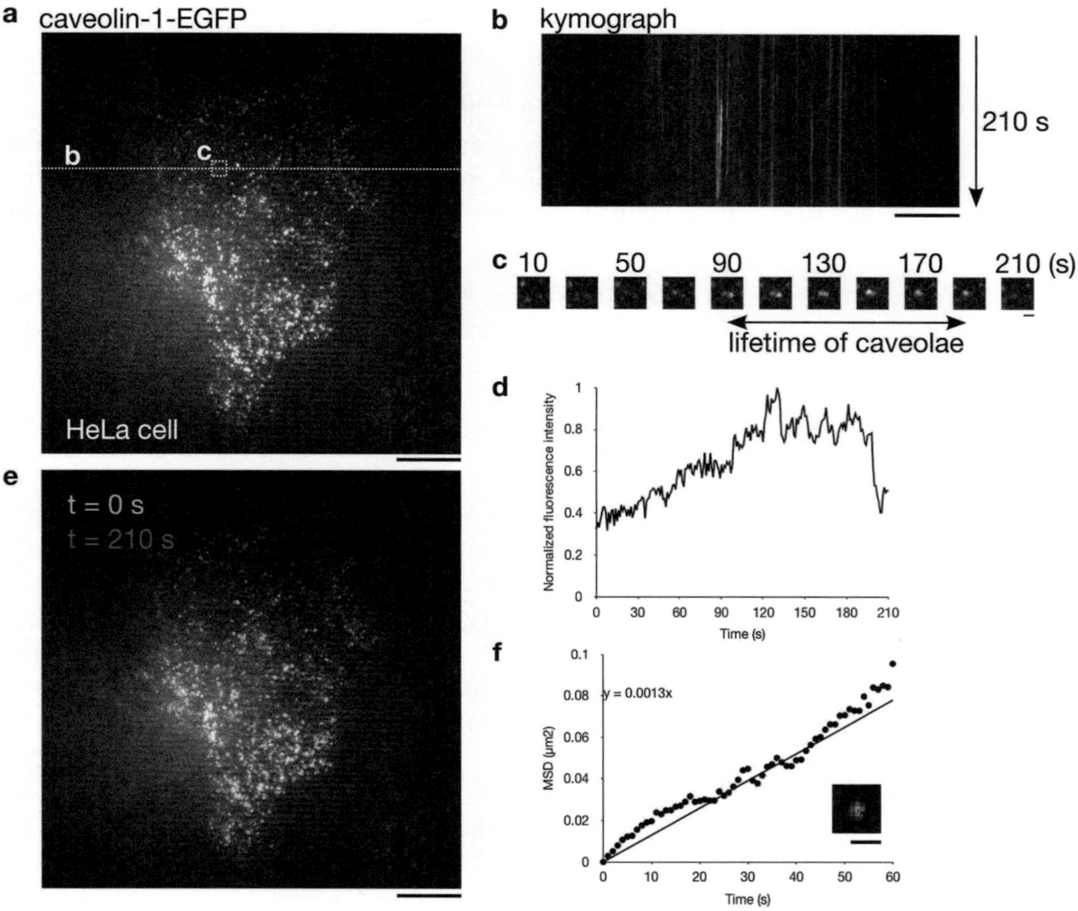

Fig. 2 Kymograph analysis of spatiotemporal caveolae dynamics imaged by total internal reflection fluorescence (TIRF) microscopy. (**a**) The distribution of caveolae on the basement plasma membrane in HeLa cell transfected with caveolin-1-EGFP under TIRF microscopy (Scale bar, 10 μm). (**b**) A representative kymograph along the dashed line drawn in the stack images (**a**) was generated using ImageJ/Fiji software. Images were acquired at 1-s intervals for 210 s. Each line represents the appearance (beginning of the vertical line) and disappearance (end of the vertical line) of individual caveolae at the basement plasma membrane. The length of the vertical line denotes caveolae lifetime (Scale bar, 10 μm). (**c**) Time-lapse TIRF images of the appearance and disappearance of individual caveolae (20-s intervals) in the ROI drawn in (**a**) (Scale bar, 1 μm) (*see* **Note 1**). (**d**) Line profile of the changes in caveolin-1-EGFP intensity in one of the caveolae in the kymograph (**c**), which was normalized to the maximum intensity. Time course of the increase and decrease in fluorescence intensity indicates caveolae appearance and disappearance at the plasma membrane, respectively. (**e**) The TIRF images of HeLa cell–expressing caveolin-1-EGFP are superimposed at $t = 0$ s (green) and $t = 210$ s (magenta), indicating that caveolae are not immobile, but moving laterally on the plasma membrane (Scale bar, 10 μm). (**f**) Lateral movements of caveolae on the plasma membrane in (**e**) were analyzed by single-particle tracking. Representative trace of the caveolae movement is displayed as magenta in the TIRF image. Diffusion coefficient D ($=3.5 \times 10^{-4}$ μm^2/s) was estimated from the slope in the mean square displacement (MSD) plots (Scale bar, 1 μm) (*see* **Note 2**)

2 Materials

2.1 Cell Culture

1. HeLa cell line (*see* **Note 3**).
2. Complete growth medium: Dulbecco's Modified Eagle Medium (DMEM) supplemented with 10% fetal bovine serum (FBS), 2 mM L-glutamine, and penicillin-streptomycin.
3. Trypsin EDTA.
4. Glass-bottom dishes: #1.5 cover glass (0.17 mm) with a low-glass-thickness variability, low auto-fluorescence, and high transmittance.
5. FuGENE HD.
6. Opti-MEM.

2.2 Microscopy

1. Nikon Instruments Eclipse Ti-E Inverted Microscope equipped with 100× objective and iXon EMCCD cameras (Andor) controlled by NIS-Elements (*see* **Note 4**).
2. Olympus FLUOVIEW FV1000 confocal laser scanning microscope equipped with UAPON100xOTIRF. NA: 1.49 (oil immersion) and Photometrics Cascade II EMCCD Camera controlled by MetaMorph (Molecular Devices).

2.3 Software

1. ImageJ/Fiji (NIH).
2. MATLAB (MathWorks).

3 Methods

3.1 Transfection

1. Culture cells in complete growth medium and maintain cells at 37 °C in 5% CO_2 incubator.
2. Trypsinize cells with trypsin EDTA, and subculture cells on glass-bottom dishes (*see* **Note 5**).
3. When the cell density reaches approximately 50–80% confluence, transfect the cells with caveolae markers such as enhanced green fluorescent protein (EGFP)–tagged caveolin-1 plasmid, using FuGENE HD diluted in Opti-MEM according to the manufacturer's protocol (*see* **Notes 6–8**).
4. Incubate the transfected cells for a further 18–72 h (*see* **Note 9**).

3.2 TIRF Microscopy

1. Before microscopy observation, add 25 mM HEPES to the cell-culture medium to maintain its pH (*see* **Note 10**). Keep microscopy at 37 °C with 5% CO_2 supply.

2. Carry out TIRF microscopy such as Nikon Instruments Eclipse Ti-E Inverted Microscope (*see* **Note 11**) or Olympus FLUOVIEW FV1000 confocal laser scanning microscope placed on an antivibration table, according to the manufacturer's instructions.
3. Select 488-nm laser line to excite EGFP-tagged proteins (*see* **Note 12**).
4. Choose the incident angle for the laser line to generate the total internal reflection, and the evanescent light selectively illuminates the basement plasma membrane of the cells with a penetration depth of ~100 nm.
5. Decrease the excitation light as much as possible to prevent photobleaching of the EGFP-tagged proteins and phototoxicity to the cells.
6. Adjust the exposure time to acquire optimal S/N images (*see* **Note 13**).
7. Acquire images at 1-s intervals for 5 min, for example.

3.3 Image Analysis

1. Analyze caveolae dynamics in the acquired time-lapse movies using ImageJ/Fiji or MATLAB software.
2. Generate a kymograph using ImageJ/Fiji to estimate the lifetime of caveolae on the plasma membrane. For example, draw a straight line along the caveolae on the basement plasma membrane in the stack images using the line ROI tool from the toolbar. Then, select the menu item "Image>Stacks>Reslice," and a kymograph will be generated along the defined line.
3. To measure the signal intensities of caveolae during their appearance/disappearance at the plasma membrane, draw a straight line along one of the caveolae traces in the kymograph images using the line ROI tool from the toolbar. Then, select the menu item "Analyze > Plot Profile." Estimate caveolae lifetime by measuring the length of the vertical traces in the kymograph.
4. Track motion of the individual caveolae on the plasma membrane using particle track plugin in ImageJ/Fiji (*see* **Note 14**). Calculate mean square displacement (MSD) from x and y coordinates in each caveolae trajectory. Plot MSD as a function of time. Estimate diffusion constant of caveolae by fitting MSD plot using the formula $y = 4 \times Dt^{\alpha}$ (normal diffusion when $\alpha = 1$, subdiffusion when $\alpha < 1$, and superdiffusion when $\alpha > 1$).

4 Notes

1. Confirm that the decrease in fluorescence intensity is due to the disappearance of endocytic machinery from the plasma membrane and not the result of photobleaching.
2. Observe several caveolae by TIRF microscopy, and lifetime, changes in fluorescence intensity, and diffusion coefficients can be quantified.
3. Cell lines were routinely tested for mycoplasma contamination.
4. Super-resolution microscopy combined with TIRF can achieve high-resolution imaging of caveolae structure in fixed cells [10].
5. The glass-bottom dishes can be coated with an extracellular matrix such as collagen, fibronectin, or laminin depending on the cell line.
6. Because fluorescent tags might alter the structural properties and interfere with the functions of caveolae markers, we validated their functions using both N-terminal and C-terminal tags to compare the localization of fluorescently tagged proteins with the known localization of endogenous proteins.
7. For difficult-to-transfect cells, an electroporation-based Neon Transfection System was used to obtain high transfection efficiency.
8. In some cases, a DsRed/mCherry/RFP-tagged caveolae marker such as Cavin [11] or dynamin2 was cotransfected for colocalization analysis. Furthermore, RNAi knockdown or CRISPR-Cas9-mediated knockout of caveolae markers is applicable.
9. We optimized the transfection efficiency (the ratio of transfection reagent to DNA, the amount of DNA used, incubation time) to control protein expression similar to endogenous levels. Alternatively, tetracycline-inducible stable cell lines can be established, or genome-editing techniques such as CRISPR-Cas9-mediated knock-in of fluorescent tags to the endogenous proteins can be applied [12, 13].
10. Phenol red–free DMEM can help to reduce the background fluorescence from the cell-culture medium.
11. The Nikon Perfect Focus System was applied during TIRF observation to prevent focus drift. This is especially useful when adding inhibitors such as dynamin GTPase (a molecule that pinches off endocytic vesicles from the plasma membrane) inhibitor dynasore [14] to the cell-culture medium during the observation.

12. When cells are cotransfected with GFP- and mCherry-tagged proteins, excite EGFP with a 488-nm laser line and mCherry/DsRed/RFP with a 561-nm laser line.
13. If the signals from the fluorescently tagged caveolae markers are too low, the electron-multiplying CCD (EMCCD) cameras with fast readout times, low noise, and high quantum efficiency (QE) instead of conventional CCD cameras might be beneficial. Alternatively, the GFP tag can be substituted with brighter variants such as super folder GFP (sfGFP) [15].
14. Several ImageJ plugins and MATLAB codes are available [16–18].

Acknowledgments

This work was supported by grants from the Funding Program for Next Generation World Leading Researchers (NEXT program) (grant number: LS031); Japan Society for the Promotion of Science (JSPS) KAKENHI (grant numbers: 26291037, 15H01641, 15H05902). Astellas Foundation for Research on Metabolic Disorders to S.S.

References

1. Sako Y, Minoghchi S, Yanagida T (2000) Single-molecule imaging of EGFR signalling on the surface of living cells. Nat Cell Biol 2:168–172
2. Boucrot E, Howes MT, Kirchhausen T, Parton RG (2011) Redistribution of caveolae during mitosis. J Cell Sci 124:1965–1972
3. Johnson A, Vert G (2017) Single event resolution of plant plasma membrane protein endocytosis by TIRF microscopy. Front Plant Sci 8:612
4. Mattheyses AL, Simon SM, Rappoport JZ (2010) Imaging with total internal reflection fluorescence microscopy for the cell biologist. J Cell Sci 123:3621–3628
5. Rosendale M, Perrais D (2017) Imaging in focus: imaging the dynamics of endocytosis. Int J Biochem Cell Biol 93:41–45
6. Senju Y, Itoh Y, Takano K et al (2011) Essential role of PACSIN2/syndapin-II in caveolae membrane sculpting. J Cell Sci 124:2032–2040
7. Senju Y, Suetsugu S (2015) Possible regulation of caveolar endocytosis and flattening by phosphorylation of F-BAR domain protein PACSIN2/Syndapin II. BioArchitecture 5:70–77
8. Sinha B, Köster D, Ruez R et al (2011) Cells respond to mechanical stress by rapid disassembly of caveolae. Cell 144:402–413
9. Senju Y, Rosenbaum E, Shah C et al (2015) Phosphorylation of PACSIN2 by protein kinase C triggers the removal of caveolae from the plasma membrane. J Cell Sci 128:2766–2780
10. Tachikawa M, Morone N, Senju Y et al (2017) Measurement of caveolin-1 densities in the cell membrane for quantification of caveolar deformation after exposure to hypotonic membrane tension. Sci Rep 7:7794
11. Lamaze C, Tardif N, Dewulf M et al (2017) The caveolae dress code: structure and signaling. Curr Opin Cell Biol 47:117–125
12. Grassart A, Cheng AT, Hong SH et al (2014) Actin and dynamin2 dynamics and interplay during clathrin-mediated endocytosis. J Cell Biol 205:721–735
13. Shvets E, Bitsikas V, Howard G et al (2015) Dynamic caveolae exclude bulk membrane proteins and are required for sorting of excess glycosphingolipids. Nat Commun 6:6867
14. Macia E, Ehrlich M, Massol R et al (2006) Dynasore, a cell-permeable inhibitor of dynamin. Dev Cell 10:839–850

15. Pédelacq J-D, Cabantous S, Tran T et al (2006) Engineering and characterization of a superfolder green fluorescent protein. Nat Biotechnol 24:79–88
16. Yoshiyuki Arai. Fast single molecule particle tracking and analysis plugin with Java Native Interface (ImageJ User and Developer Conference). http://imagejconf.tudor.lu/program/poster/yoshiyuki_arai73473173
17. Smith MB, Karatekin E, Gohlke A et al (2011) Interactive, computer-assisted tracking of speckle trajectories in fluorescence microscopy: application to actin polymerization and membrane fusion. Biophys J 101:1794–1804
18. Jaqaman K, Loerke D, Mettlen M et al (2008) Robust single-particle tracking in live-cell time-lapse sequences. Nat Methods 5:695–702

Chapter 7

Live-Cell FRET Imaging of Phosphorylation-Dependent Caveolin-1 Switch

Adriana M. Zimnicka, Zhenlong Chen, Peter T. Toth, and Richard D. Minshall

Abstract

The detection of dynamic conformational changes in proteins in live cells is challenging. Live-cell FRET (Förster Resonance Energy Transfer) is an example of a noninvasive technique that can be used to achieve this goal at nanometer resolution. FRET-based assays are dependent on the presence of fluorescent probes, such as CFP- and YFP-conjugated protein pairs. Here, we describe an experimental protocol in which live-cell FRET was used to measure conformational changes in caveolin-1 (Cav-1) oligomers on the surface of plasmalemma vesicles, or caveolae.

Key words Live-cell FRET, Cav-1-CFP, Cav-1-YFP, Caveolae, Conformation change

1 Introduction

Formation and function of caveolae require caveolin-1 (Cav-1) and cavin-family proteins inserted or associated with cholesterol and phospholipid-enriched microdomains of the plasma membrane [1–3]. Caveolae play critical roles in a variety of cellular functions, including endocytosis and transcytosis [4], the establishment of signaling hubs [5], storage of cholesterol [6], and buffering of mechanical stress [7]. The caveolar coat is far from static [8], but uncovering the intricate mechanisms by which Cav-1 molecules rearrange themselves to form tighter or looser associations upon receiving relevant cellular stimuli requires monitoring at the nanometer scale. One of the methods capable of reaching such nanometer resolution is FRET, which depends on fluorescence-based probes directly conjugated to proteins and biosensors. The most commonly used fluorescent probe is green fluorescent protein or GFP, which was originally derived from the jellyfish, *Aequorea victoria* [9]. Since then, the original GFP protein has undergone numerous mutations to produce enhanced GFP (eGFP), blue

Cedric M. Blouin (ed.), *Caveolae: Methods and Protocols*, Methods in Molecular Biology, vol. 2169,
https://doi.org/10.1007/978-1-0716-0732-9_7,

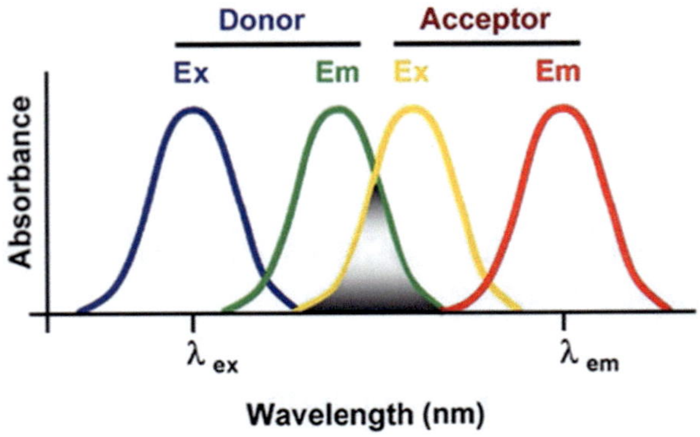

Fig. 1 Idealized absorption and emission spectra for FRET donor and acceptor pair, such as ECFP (donor) and EYFP (acceptor) fluorescent proteins. Absorption spectra are illustrated as blue and yellow curves, while the emission spectra are presented as green and red curves. The region of overlap between the donor emission and acceptor absorption spectrum is represented by a gray area. For FRET to occur, there should be a direct excitation of the acceptor by the emission spectrum of the donor, and the donor and acceptor molecules need to have the appropriate dipole alignment and be positioned within 10 nm of each other

fluorescent protein (BFP), cyan fluorescent protein (CFP), and yellow fluorescent protein (YFP) (*see* **Note 1**). Variants of enhanced CFP and YFP (ECFP and EYFP) have become the most popular FRET pairs (*see* **Note 2**).

FRET occurs in the range of 1–10 nm, due to collision-free, resonance-based energy transfer between chromophore probes, without transmission of photons [10], when the emission band of the donor exhibits spectral overlap with the absorption band of the acceptor (Fig. 1). FRET has been used extensively to study conformational changes in molecules, *cis*- and *trans*-conversions, and the assembly of macromolecules [11–13]. Traditional approaches to FRET, such as acceptor photobleaching to quantify the donor fluorescence before and after photodestruction of the acceptor, requires the sample to be exposed to high-intensity light, often resulting in cellular phototoxicity [14]. In contrast, the live FRET technique, described later, assesses FRET efficiency from brief low-intensity light pulses, enabling rapid and dynamic measurements with relatively minimal photodamage in living cells (*see* **Note 3**). Live FRET microscopy can be employed to assess protein-protein interactions, such as those exhibited by Cav-1 monomers in their natural environment, and to measure spatial and temporal changes as they are occurring between binding partners in a single living cell. We have used this nondestructive technique to study changes in protein-protein interactions at the caveolar coat due to phosphorylation of Cav-1 at tyrosine 14 (Y14) by Src in the process of endocytic uptake of albumin in live cells [8, 15].

2 Materials

2.1 Plasmids and Primers

1. Cav-1 gene (NM_001753) (Addgene, plasmid #55468)
2. pAMCyan1-N1 (Takara, #632440) and pEYFP-N1 (Clontech #6085-1) vectors,
3. Primer pair: Cav1-WT-Fwd: 5′-ACTAGCTAGCGACCGCCA TGTCTGGGGGCAAA-TAC-3′ and
   Cav-1-WT-Rvrs: 5′-ACTGGGTACCGTTATTTCTTTCTGC AAGTTGATGCG-3′

2.2 General DNA Manipulation

1. DNA Phusion High-Fidelity Polymerase.
2. Nhe1 and Kpn1 restriction enzymes.
3. T4 DNA ligase.
4. Chemically competent cells, such as MAX Efficiency DH5α T1 (Invitrogen).
5. Agarose gel, electrophoresis apparatus for agarose gels, agar plates, Mini-Prep and Midi-Prep kits.

2.3 Generation of Cell Line Stably Expressing Fluo-Tagged Cav1

1. Lipofectamine 3000 (Invitrogen).
2. G418 or Geneticin.

2.4 FRET Picture Acquisition

1. Glass-bottom dishes (MatTek).
2. Confocal microscope such as Zeiss 710 BIG (Carl Zeiss).

3 Methods

3.1 Labeling of the Cav-1 Protein at the C-Terminus with EYFP or ECFP Fluorescent Protein Tags

1. Combine the primer pair lacking the stop codon: Cav1-WT-Fwd Cav-1-WT-Rvrs: with the full-length Homo sapiens Cav-1 gene as a template in a PCR reaction using DNA Phusion High-Fidelity Polymerase according to the manufacturer's instructions [8, 15].
2. Purify the resulting PCR fragment by running agarose gel.
3. Digest the purified PCR fragment, as well as pEYFP-N1 vector, with restriction enzymes 5′-NheI and 3′-KpnI (restriction sites underlined in the primers) according to the manufacturer's instructions.
4. Ligate digested PCR fragment with a digested pEYFP-N1 vector using T4 DNA ligase, according to the manufacturer's instructions.
5. Subsequently, transform DH5α bacteria with ligated DNA and spread them on agar plates.

6. Isolate DNA from single colonies on the agar plate and purify them using Mini-prep kit.
7. Verify the correct insertion of the fluorescent tags by Sanger sequencing using standard CMV fwd sequencing primer. Prepare Midi-prep of the correct DNA for future transfections.
8. Generate the Cav-1-CFP construct by repeating **steps 1–7** but by substituting pAmCyan vector for pEYFP-N1.

3.2 HEK Cell Transfections

1. Split and culture HEK cells on 6-cm dishes for 2 days until they reach ~50–80% confluence (*see* **Note 4**). Transfect cells with either 2 or 4 μg of Cav-1-EYFP and Cav-1-ECFP cDNA (4 μg alone or a combination of 2 μg each) using Lipofectamine 3000 according to the manufacturer's protocol.
2. After 24 h, add G418 (geneticin) to the cells at a concentration of 0.5 mg/mL. Stable expression of cDNA in HEK cells should be obtained within 10–14 days.
3. As a negative experimental control, and for determination of optimal conditions for minimizing photobleaching of fluorophores with time, cotransfect one batch of HEK cells with a combination of pEYFP-N1/pAmCyan empty vectors by repeating **steps 1–2**.
4. To measure DER and AER constants (*see* **Note 3**), prepare a batch of HEK cells transfected with Cav-1-EYFP construct only (an acceptor-only control), and a nd batch transfected with Cav-1-ECFP construct only (a donor-only control), following **steps 1** and **2**.

3.3 Plating Cells for the Experiment

1. Confirm equivalent coexpression of EYFP- and ECFP-tagged Cav-1 constructs in HEK cells intended for FRET by using fluorescent microscopy.
2. Plate four batches of HEK cells (FRET sample, a negative control, a donor-only control, and an acceptor-only control) on glass-bottom dishes and keep them growing on glass for 48 h prior to the experiment (*see* **Note 5**).
3. Deprive HEK cells of serum by changing medium to phenol-red free DMEM for 2 h before the beginning of Live FRET measurements.

3.4 Microscope Measurements

1. During long-term live-cell imaging experiments, aim to maximize the photon collection and minimize the laser power of the confocal microscope, such as Zeiss 710 BIG.
2. Use Plan Apochromat objective, at least ×63 magnification with numerical aperture 1.40 or higher.
3. Open the pinhole to about 3.3 Airy units to collect more photons, and to have optical sections of ~2 μm, so it is less

Table 1
List of excitation and emission wavelengths needed to convert raw FRET reading to a corrected FRET (cFRET)

	Sample	Excitation (nm)	Emission (nm)
I^{da}	Donor-only control	458	526–608
I^{dd}	Donor-only control	458	463–516
I^{da*}	Acceptor-only control	458	526–608
I^{aa*}	Acceptor-only control	514	526–608
I^{DA}	FRET sample	458	526–608
I^{DD}	FRET sample	458	463–516
I^{AA}	FRET sample	514	526–608

The donor spectral bleed-through and acceptor crossexcitation are accounted for, according to the formula: $\text{cFRET} = I^{DA} - \text{DER} * I^{DD} - \text{AER} * I^{AA}$, where $\text{DER} = \frac{I^{da}}{I^{dd}}$, and $\text{AER} = \frac{I^{da*}}{I^{aa*}}$

likely that the caveolin-positive vesicle leaves this optical section during the course of the experiment.

4. Use low laser power, in the 1–4% range.
5. Before the beginning of the experiment, collect fluorescence (I^{da} and I^{dd}, followed by I^{da*} and I^{aa*}; *see* **Note 3**) for the two control samples: HEK cells transfected with Cav-1-ECFP (donor only) and HEK cells transfected with Cav-1-EYFP (acceptor only) to calculate DER and AER constants. Use the absorption/emission settings described in Table 1.
6. Begin the actual FRET experiment by equilibrating HEK cells cotransfected with a mixture of Cav-1-EYFP/Cav-1-ECFP for 1 h in the environmental PeCon-heated chamber with CO_2 controller attached to the microscope stage.
7. Generate images (for ~5 min) to obtain basal fluorescence before the addition of the stimulus from ROIs (Regions of Interest) situated in the peripheral areas of the cell. Use the 458-nm laser line to excite and collect simultaneously the emission spectra in 526–608-nm region for FRET (I^{DA}), and 463–516-nm region for CFP (I^{DD}). In addition, at each time point, measure I^{AA} by exciting the FRET sample with 514-nm laser line and collecting emission fluorescence in the 526–608-nm range.
8. After ~5 min, stimulate cells with BSA (3 mg/mL; *see* **Note 6**) and then continue the measurements at each time point for FRET, CFP, and I^{AA} for an additional 20–30 min.
9. Repeat **steps 2–4** for the negative control using HEK cells cotransfected with the mixture of pEYFP-N1/pAMCyan empty vectors.

3.5 Analysis

1. Use ImageJ (Fiji) to quantify fluorescence intensities for I^{da}, I^{dd}, and I^{da*}, I^{aa*}, as well as FRET (I^{DA}), cyan (I^{DD}), and I^{AA}, in the following way:
   (a) Download and install Fiji from https://imagej.net/Fiji/Downloads
   (b) Go to Fiji analysis→ set measurements→ then, check: mean gray value, standard deviation, and area
   (c) Draw a rectangle around the area whose fluorescence you want to measure using the rectangle tool
   (d) Go to Analyze → measure, or use command "ctrl-M"
   (e) Express Intensity as a mean gray value ± standard deviation
2. First, calculate coefficients DER and AER according to the formula in **Note 3**.
3. Next, for each time point, calculate corrected FRET (cFRET) by using DER and AER constants obtained earlier:

$$\text{cFRETt} = I^{DA} - \text{DER} * \text{I}^{DD} - \text{AER} * \text{I}^{AA}$$

4. To account for temporal fluctuations in donor fluorescence, express changes in cFRET at each time point as the FRET index by dividing cFRET intensity at each time point by CFP (I^{DD}) intensity at a given time point: FRET index $= \frac{\text{cFRET}}{I^{DD}}$.
5. Subsequently, normalize to 1 the FRET index before BSA stimulation, and represent any changes in FRET after BSA stimulation relative to it (Fig. 2; *see* **Note 7**).

4 Notes

1. It is known that many fluorescent proteins (FPs) can easily form oligomers at high concentrations (mM), which may lead to unforeseen artifacts. For example, noncovalent dimerization of GFP/eGFP is a common [10] occurrence [16, 17]. Other FPs can form even higher-order oligomeric structures, such as dsRed, which has the propensity for tetramerization [18]. Many of these wild-type FPs, however, can be engineered into monomers only, such as by the introduction of the A206K mutation on the dimerization interface of the GFP protein [19, 20]. For typical FRET studies, the standard EYFP/ECFP pairs are used with the assumption that their concentration in the cell is less than the threshold needed for oligomerization [10].

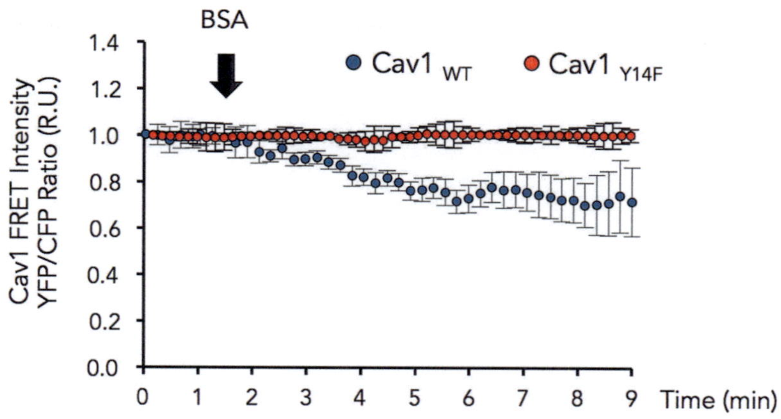

Fig. 2 Dynamic changes in caveolin-1 upon BSA treatment, monitored in real time by FRET. HEK cells were cotransfected with either wild-type or Y14F Cav1-YFP and Cav1-CFP constructs to create FRET pairs, followed by selection with G418 for stable expression. Prior to the experiment, cells were serum deprived for 2 h, and images were then acquired with a Zeiss 710 BIG confocal microscope in a temperature- and CO_2/humidity-controlled chamber. FRET was recorded every 15 s using the 458-nm laser line for excitation; the 526- to 608-nm emission range was measured for FRET and the 463- to 516-nm emission range for CFP, recorded as an internal control. After temperature equilibration and stabilization of fluorescence, recordings began and basal FRET was collected for 2–5 min before stimulation with BSA (3 mg/mL), after which FRET was collected for an additional 9 min. At each time point, the FRET index was calculated (FRET/CFP) to account for temporal fluctuations in fluorescence. The FRET index before BSA stimulation (basal) was normalized to 1, and all subsequent time points were expressed as a fraction of the basal fluorescence. The FRET trace is shown for WT-Cav-1 and Y14F-Cav1. Data are shown as mean $\pm$ SD

2. The sensitivity of biosensors based on ECFP and EYFP is generally limited [21]. A high-efficiency FRET pair, CyPet and YPet, has been developed to significantly enhance the dynamic range of FRET [22]. However, CyPet is not suitable for live-cell imaging as it folds very poorly at 37 °C [23].
3. This method [24] requires correction for crosstalk, which consists of the donor spectral bleed-through (SBT) and acceptor cross excitation. Crosstalk is corrected by two reference measurements, in addition to the actual FRET experiment [24]. The first is performed using a control sample containing only donor fluorophores in order to account for donor bleed through (DER). The second is performed on a sample containing only acceptor fluorophores to account for direct acceptor excitation (AER). FRET can be corrected for the crosstalk according to the following formula:

$$\text{cFRET} = I^{\text{DA}} - \text{DER} * \text{I}^{\text{DD}} - \text{AER} * \text{I}^{\text{AA}}, \quad \text{where}$$

DER $= \frac{I^{da}}{I^{dd}}$, measured from donor-only control sample;

AER $= \frac{I^{da*}}{I^{aa*}}$, measured from acceptor-only control sample;

while I^{DA}, I^{DD}, and I^{AA} are all measured from the actual FRET sample, containing the donor/acceptor pair. To calculate DER, the donor-only sample is illuminated with donor excitation and fluorescence intensity is collected at acceptor emission (I^{da}); next, the donor-only sample is illuminated with donor excitation and fluorescence is collected at donor emission (I^{dd}).

To calculate AER, the acceptor-only sample is illuminated with donor excitation and fluorescence intensity is collected at acceptor emission (I^{da*}); next, the acceptor-only sample is illuminated with acceptor excitation and fluorescence intensity is collected at acceptor emission (I^{aa*}).

To measure I^{DA}, the fluorescence intensity is obtained with donor excitation and collected at acceptor emission in FRET sample; I^{DD} fluorescence intensity is obtained with donor excitation collected at donor emission in FRET sample; and finally, I^{AA} fluorescence intensity is obtained with acceptor excitation collected at acceptor emission in FRET sample.

4. All cell experiments reported here were performed in HEK cells. Cells were cultured in Dulbecco's modified Eagle's medium (DMEM) supplemented with 10% fetal bovine serum (FBS), 10 mM HEPES, 1% penicillin/streptomycin (10,000 U/mL), and, as required to maintain stable expression, G418 (Geneticin). Cells were cultured at 37 °C and 5% CO_2 and maintained in 10-cm tissue culture dishes.

5. You may want to culture cells on poly-D-lysine–treated glass-bottom dishes, if the attachment of cells to the glass surface is a problem. Prepare 1 mg/mL stock solution by dissolving 100 mg Poly-D-lysine hydrobromide in 100-mL water. Filter sterilize the solution, then store aliquots at −20 °C. The working solution is obtained by diluting the stock solution with sterile water to 0.1 mg/mL. The working solution is then added to the dish in such a way that the glass surface is completely covered and incubated for 1 h at room temperature in the hood. Subsequently, the poly-D-lysine solution is suctioned off and dishes are washed twice in PBS.

6. Addition of BSA was performed to the side of the dish via tubing connected to a syringe, without opening of the environmental chamber, with as little disruption to the cells as possible.

7. Figure 2 depicts FRET measured continuously in live HEK cells before and after stimulation with 3 mg/mL BSA. HEK cells were cotransfected with YFP- and CFP-conjugated Cav1 cDNA coding for either wild-type Cav1, or Y14F mutant.

Within 10 min of BSA stimulation, there is a decrease in FRET signal (Cav1-YFP/Cav-1-CFP ratio) to about ~60% of the initial value for WT Cav-1. No significant changes in FRET were observed when the irreversible, phosphodefective Y14F mutant was stimulated with BSA. This indicates that phosphorylation-dependent changes at tyrosine 14 introduced by mutation trump any changes in caveolar coat packing due to phosphorylation.

References

1. Fernandez I, Ying Y, Albanesi J, Anderson RG (2002) Mechanism of caveolin filament assembly. Proc Natl Acad Sci U S A 99:11193–11198
2. Shvets E, Ludwig A, Nichols BJ (2014) News from the caves: update on the structure and function of caveolae. Curr Opin Cell Biol 29:99–106
3. Parton RG, Simons K (2007) The multiple faces of caveolae. Nat Rev Mol Cell Biol 8:185–194
4. Sverdlov M, Shinin V, Place AT, Castellon M, Minshall RD (2009) Filamin A regulates caveolae internalization and trafficking in endothelial cells. Mol Biol Cell 20:4531–4540
5. Lisanti MP, Tang Z, Scherer PE, Kubler E, Koleske AJ, Sargiacomo M (1995) Caveolae, transmembrane signalling and cellular transformation. Mol Membr Biol 12:121–124
6. Fielding CJ, Fielding PE (2001) Caveolae and intracellular trafficking of cholesterol. Adv Drug Deliv Rev 49:251–264
7. Sinha B, Koster D, Ruez R, Gonnord P, Bastiani M, Abankwa D, Stan RV, Butler-Browne G, Vedie B, Johannes L, Morone N, Parton RG, Raposo G, Sens P, Lamaze C, Nassoy P (2011) Cells respond to mechanical stress by rapid disassembly of caveolae. Cell 144:402–413
8. Zimnicka AM, Husain YS, Shajahan AN, Sverdlov M, Chaga O, Chen Z, Toth PT, Klomp J, Karginov AV, Tiruppathi C, Malik AB, Minshall RD (2016) Src-dependent phosphorylation of caveolin-1 Tyr-14 promotes swelling and release of caveolae. Mol Biol Cell 27:2090–2106
9. Shimomura O, Johnson FH, Saiga Y (1962) Extraction, purification and properties of aequorin, a bioluminescent protein from the luminous hydromedusan, Aequorea. J Cell Comp Physiol 59:223–239
10. Shrestha D, Jenei A, Nagy P, Vereb G, Szollosi J (2015) Understanding FRET as a research tool for cellular studies. Int J Mol Sci 16:6718–6756
11. Gauer JW, LeBlanc S, Hao P, Qiu R, Case BC, Sakato M, Hingorani MM, Erie DA, Weninger KR (2016) Single-molecule FRET to measure conformational dynamics of DNA mismatch repair proteins. Methods Enzymol 581:285–315
12. Katz RA, Merkel G, Andrake MD, Roder H, Skalka AM (2011) Retroviral integrases promote fraying of viral DNA ends. J Biol Chem 286:25710–25718
13. Ben-Johny M, Yue DN, Yue DT (2016) Detecting stoichiometry of macromolecular complexes in live cells using FRET. Nat Commun 7:13709
14. Butz ES, Ben-Johny M, Shen M, Yang PS, Sang L, Biel M, Yue DT, Wahl-Schott C (2016) Quantifying macromolecular interactions in living cells using FRET two-hybrid assays. Nat Protoc 11:2470–2498
15. Chen Z, Bakhshi FR, Shajahan AN, Sharma T, Mao M, Trane A, Bernatchez P, van Nieuw Amerongen GP, Bonini MG, Skidgel RA, Malik AB, Minshall RD (2012) Nitric oxide-dependent Src activation and resultant caveolin-1 phosphorylation promote eNOS/caveolin-1 binding and eNOS inhibition. Mol Biol Cell 23:1388–1398
16. Snapp EL, Hegde RS, Francolini M, Lombardo F, Colombo S, Pedrazzini E, Borgese N, Lippincott-Schwartz J (2003) Formation of stacked ER cisternae by low affinity protein interactions. J Cell Biol 163:257–269
17. Tsien RY (1998) The green fluorescent protein. Annu Rev Biochem 67:509–544
18. Matz MV, Fradkov AF, Labas YA, Savitsky AP, Zaraisky AG, Markelov ML, Lukyanov SA (1999) Fluorescent proteins from

nonbioluminescent Anthozoa species. Nat Biotechnol 17:969–973

19. von Stetten D, Noirclerc-Savoye M, Goedhart J, Gadella TW Jr, Royant A (2012) Structure of a fluorescent protein from Aequorea victoria bearing the obligate-monomer mutation A206K. Acta Crystallogr Sect F Struct Biol Cryst Commun 68:878–882
20. Zacharias DA, Violin JD, Newton AC, Tsien RY (2002) Partitioning of lipid-modified monomeric GFPs into membrane microdomains of live cells. Science 296:913–916
21. Ouyang M, Sun J, Chien S, Wang Y (2008) Determination of hierarchical relationship of Src and Rac at subcellular locations with FRET biosensors. Proc Natl Acad Sci 105:14353–14358
22. Nguyen AW, Daugherty PS (2005) Evolutionary optimization of fluorescent proteins for intracellular FRET. Nat Biotechnol 23:355–360
23. Shaner NC, Steinbach PA, Tsien RY (2005) A guide to choosing fluorescent proteins. Nat Methods 2:905–909
24. Elder AD, Domin A, Kaminski Schierle GS, Lindon C, Pines J, Esposito A, Kaminski CF (2009) A quantitative protocol for dynamic measurements of protein interactions by Förster resonance energy transfer-sensitized fluorescence emission. J R Soc Interface 6:S59–S81

Chapter 8

GPMVs as a Tool to Study Caveolin-Interacting Partners

Joanna Podkalicka and Cedric M. Blouin

Abstract

Caveolins, major components of small plasma membrane invaginations called caveolae, play a role in signaling, particularly in mechanosignaling. These proteins are known to interact with a variety of effector molecules, including G-protein-coupled receptors, Src family kinases, ion channels, endothelial nitric oxide synthase (eNOS), adenylyl cyclases, protein kinase A (PKA), and mitogen-activated PKs (MAPKs). There is, however, speculation on the relevance of these interactions and the mechanisms by which caveolins may control intracellular signaling. This chapter introduces a method of isolation of giant plasma membrane-derived vesicles (GPMVs), which possess full complexity of membrane they originate from, thus comprising an excellent platform to revisit some of the previously described interactions in a cleaner environment and possibly identifying new binding partners. It is also a powerful technique for studying membrane mechanics, as it was previously used to demonstrate the role of caveolae in mechanoprotection.

Key words Caveolin, GPMV, Model membrane, Mechanosignaling, Protein-protein interaction

1 Introduction

Giant plasma membrane-derived vesicles (GPMVs) comprise a model membrane system, which retains the lipid/protein complexity of the plasma membrane (PM). They are chemically induced PM vesicles (PMVs) devoid of internal membranous structures and assembled cytoskeleton components [1]. Although this type of membrane vesiculation was first observed in 1961 [2], PMVs were not commonly used as a model system due to the chemical modifications induced during their isolation. The fate-changing moment for the method came with the observation of liquid–liquid phase separation in GPMVs described by the group of Baumgart in 2007 [3]. Since then, numerous studies have focused on phase separation property of these vesicles to establish partitioning of different proteins between liquid-disordered (ld) and liquid-ordered (lo) phases of the membrane ([4–6], for review *see* [7]). Many methodologic advancements were made to minimize the chemical modifications of membrane components, leading to the introduction of vesiculation agents such as *N*-ethyl maleimide

Cedric M. Blouin (ed.), *Caveolae: Methods and Protocols*, Methods in Molecular Biology, vol. 2169,
https://doi.org/10.1007/978-1-0716-0732-9_8,

(NEM), which irreversibly reacts with terminal sulfhydryl moieties (typically cysteine side chains), covalently blocking these groups without inducing any crosslinking of proteins or lipids [8, 9]. Nowadays with optimized protocols for GPMV isolation, the technique represents an excellent tool to study membrane properties in a system, which fully maintains PM complexity giving insight into physicochemical properties of the pure PM. The technique is suitable for all types of microscopy observations as well as a wide variety of biochemical assays (for review and exhaustive protocol with applications, *see* Sezgin et al. [10]).

Caveolae are protein-mediated, SM-cholesterol-enriched invaginations of the PM of 50–80 nm in diameter [11]. They are important in a variety of cellular processes, including lipid homeostasis and cellular signaling in physiological as well as in pathological conditions. Caveolae formation is dependent on two among three isoforms of caveolin proteins: Cav1 or Cav3 depending on tissue [12, 13] and cytosolic cavins. Caveolins are 18–22 kDa integral membrane proteins, which insert asymmetrically in a hairpin-like conformation. Caveolae were shown to play a major role in mechanoprotection and mechanosignaling by flattening out under mechanical stress [14, 15]. Caveolin-dependent signaling is thought to be mediated through caveolin-scaffolding domain (CSD, residues 82–101) and a putative caveolin-binding motif (CBM) within interactor protein. CSD would bind several signaling effectors, including some growth factor receptors, eNOS, Src tyrosine kinases, G-proteins, and G-protein-coupled receptors [16, 17]; however, CBM relevance for caveolin binding is still under debate due to its structural analysis [18].

GPMV isolation could constitute an interesting tool to identify caveolin-binding partners (in caveolae) as it provides pure PM environment without contaminations from ER and Golgi caveolin pools. It has been previously shown that GPMVs could be successfully used for caveolin/cavin colocalization studies as well as membrane mechanics while combined with micropipette aspiration methods [14].

2 Materials

2.1 Cell Culture

Retinal Pigment Epithelium (RPE1) cells are maintained in Dulbecco's Modified Eagle Medium/F-12 (DMEM/F-12) supplemented with 10% fetal calf serum (FCS) and Penicillin/Streptomycin at 37 °C in a humidified atmosphere of 5% CO_2. For GPMV isolation, cells are seeded the day before at the density allowing around 70–80% confluence on the day of the experiment. Typically for microscopy observation of GPMVs 12- or 6-well plate is sufficient, whereas for biochemical analysis at least two 10-cm dishes are required (depending on cell type and the yield of GPMV generation—*see* **Note 1**). Optional: manipulate cells before GPMV isolation depending on your experiment (e.g., transfection with fluorescently tagged proteins, which are supposed to be analyzed).

2.2 GPMV Isolation

1. GPMV buffer: 10 mM HEPES, 150 mM NaCl, 2 mM $CaCl_2$ in double distilled (ddH_2O). Adjust to pH 7.4 with NaOH. Filtered buffer can be stored at room temperature for 6 months.
2. *N*-ethyl maleimide (NEM): 1 M stock solution in ethanol. Caution: NEM is toxic—use gloves (*see* **Note 2** for alternative vesiculation agent). NEM stock solution can be stored at −20 °C for a few months. Avoid using once the solution turns yellow.
3. Vesiculation buffer: GPMV buffer supplemented with 10 mM NEM and 2% (w/v) EDTA-free complete proteases inhibitor cocktail. Prepare fresh prior to the experiment.
4. PBS: 137 mM NaCl, 2.7 mM KCl, 10 mM $Na_2HPO_4{\cdot}2H_2O$, 2 mM KH_2PO_4 in ddH_2O. Adjust to pH 7.4 with HCl.

2.3 GPMV Imaging

1. 0.1% β-casein in ddH_2O. Store up to 1 year at −20 °C.
2. PBS
3. 22 × 22 mm and 24 × 60 mm coverslips or LabTek chambers.
4. FAST-DiO solution: 3,3′-Dilinoleyloxacarbocyanine perchlorate powder is dissolved in ethanol at 0.5 mg/mL. The stock solution should be kept up to 1 year at −20 °C protected from light.
5. Coverslip sealant: either paraffin wax, technical Vaseline, silicone, or high-vacuum grease can be used.

2.4 GPMV Biochemical Analysis

1. Resuspension buffer: PBS supplemented with 2% (w/v) EDTA-free complete proteases inhibitor cocktail
2. Bradford solution

3 Methods

3.1 GPMV Isolation

1. Prepare fresh vesiculation buffer just before the GPMV isolation.
2. Wash cells once with PBS and subsequently once with GPMV buffer.
3. Add 1 mL of vesiculation buffer for single well of 6-well plate (for 10-cm dish use 4 mL of vesiculation buffer, for more tips *see* **Note 1**).
4. Incubate cells for 1 h at 37 °C.

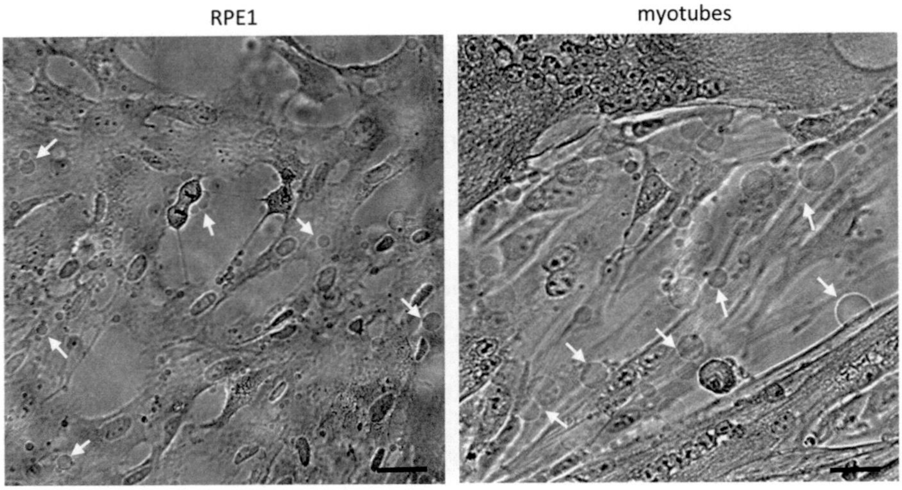

Fig. 1 Bright-field microscopy images of RPE1 cells and human myotubes after 1 h of NEM-induced GPMV formation. RPE1 produce a high amount of relatively small GPMVs, whereas myotubes produce large GPMVs in high quantities. Exemplary GPMVs marked with white arrows. Scale bar 5 μm

5. Check the presence of GPMVs in a standard bright-field microscope with 20× magnification. Vesicles should be visible as dark spheres floating at the plane of cells (Fig. 1) (*see* **Note 1**).
6. Carefully decant GPMV-containing supernatant to an Eppendorf or Falcon tube (depending on the volume). Do not pipette up and down to avoid detachment of the cells (*see* **Note 1**). From this step, proceed either with imaging (Subheading 3.2) or biochemical analysis (Subheading 3.3)

3.2 GPMV Imaging

1. Preparation of coverslips:
   (a) Passivate 22 × 22 mm coverslips or LabTek chambers by incubating with 0.1% β-casein for 1 h at room temperature. Passivation should be prepared just before imaging.
   (b) Wash coverslips/LabTek chambers two times with PBS. Keep in PBS before GPMV imaging.
2. Imaging:
   (a) Let the GPMVs sediment in the Eppendorf tube by leaving collected suspension at 4 °C for 30 min.
   (b) Pipette carefully 100 μL of GPMV suspension from the **bottom of the tube** and transfer to the new one for membrane labeling if required (if no membrane labeling is performed, proceed directly to point d).
   (c) Add 1 μL of Fast-DiO stock solution to GPMVs. Mix gently and incubate for 15 min at room temperature protected from light.

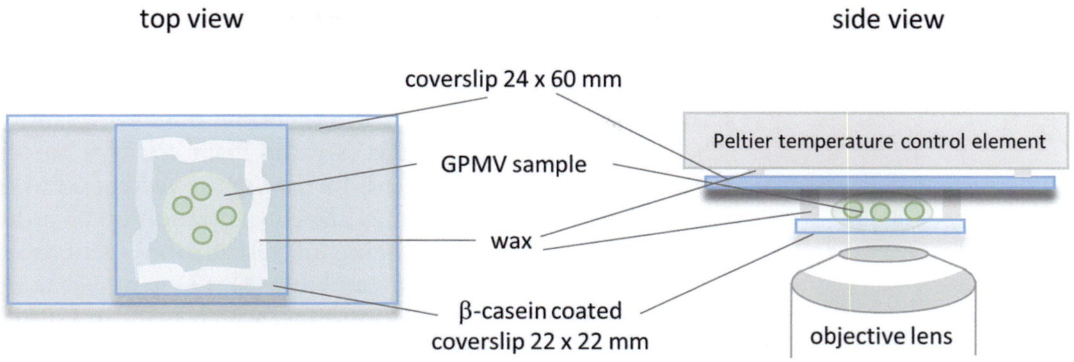

Fig. 2 Schematic representation of imaging chamber constructed from two coverslips (for advantage of using such a chamber, *see* **Note 3**)

(d) If using a coverslip chamber, put the wax sealant with a pipette tip in the square shape of approximately 15 × 15 mm on the top of the 24 × 60 mm coverslip. Pipette 15–20 μL of GPMV suspension into the middle of the square. Seal with β-casein passivated 22 × 22 mm coverslip and turn the chamber upside down (Fig. 2) (*see* **Note 3**).

If using a LabTek chamber, add 100 μL of GPMV buffer per well and subsequently add 20–50 μL of GPMV suspension.

(e) Incubate for 5–10 min to let GPMVs sediment.

(f) Image using standard microscopy methods (*see* **Note 3**).

3.3 GPMV Biochemical Analysis

1. Centrifuge GPMV suspension at 200 × *g* for 5 min at 4 °C to remove cell debris. Transfer the supernatant carefully into ultracentrifugation tube.
2. Centrifuge GPMV suspension at 100,000 × *g* for 1 h at 4 °C. Discard the supernatant.
3. Resuspend the GPMV membrane pellet in 50 μL of resuspension buffer.
4. Measure the total concentration of protein using Bradford assay (BCA is suitable too).
5. Use GPMV membranes for standard biochemical assays like Western Blot or coimmunoprecipitation.

4 Notes

1. The GPMV yield and the quality of the preparation depend on numerous parameters. In principle, the higher the cell number used, the more GPMVs in the final preparation. However, when cells are too confluent, it can result in a decreased number

of vesicles. Usually, 70–80% cell confluency gives the highest yield of vesicles. For imaging, it is usually sufficient to isolate GPMVs from a well of 12- or 6-well plate, whereas for biochemical analysis 10-cm dish is usually the minimal quantity. The number of isolated GPMVs varies from cell type to cell type and needs to be optimized. Increasing the yield of GPMVs is possible by (1) varying the NEM concentration up to order of magnitude [9], (2) changing of vesiculation agent from NEM to PFA-DTT mixture (*see* **Note 2**), (3) gentle shaking of cell plates during vesiculation time, which helps in GPMV detachment from cell membrane resulting in higher yield collected with the supernatant. However, shaking could also increase the risk of cell detachment, which will result in lower purity of isolated GPMVs. Whereas for imaging cellular contamination is not a huge problem since it is easy to differentiate GPMVs from cells and cell debris under the microscope, for biochemical analysis any cellular contamination should be avoided. For that reason, the collection of GPMV suspension from above the cells needs to be performed carefully without repetitive pipetting. One-time gentle aspiration should be sufficient to collect most of the vesicles while cells still remain attached to the dish.

2. Instead of NEM, another vesiculation agent can be used with the addition of 25 mM PFA/2 mM DTT to the GPMV buffer. The solution should be freshly prepared. It is the most efficient vesiculation agent, but it creates some artifacts like crosslinking of lipids and proteins. This can make this preparation not suitable for further biochemical analysis because of reduction of thioester bonds and disulfide bridges, coupling of phosphatidylethanolamines to proteins. All of these modifications need to be taken into account while analyzing data from GPMVs obtained with this method. This is why the use of NEM, which is a non-cross-linking vesiculant, is highly recommended. Unfortunately, NEM often gives lower yields of GPMVs.

3. GPMVs are easily observable by standard microscopy methods, ranging from bright field, EPI fluorescence, to confocal and two-photon microscopy. They found a great use in phase separation studies showing partitioning of proteins to either liquid-disordered (ld) or liquid-ordered (lo) phases. To obtain phase separation in NEM-generated GPMVs, cooling of the sample is required. For this application, the proposed imaging chamber (Fig. 2) is of great use as it enables easy coupling to a Peltier temperature controlling unit where almost all the surface of the coverslip is in close contact with the cooling plate, though providing even temperature distribution within the sample. Moreover, the proposed chamber is suitable for small volumes of nondiluted GPMVs and it prevents it from evaporation while

properly sealed. For more details on phase separation studies using GPMVs, *see* work of Sezgin and colleagues [10].

Acknowledgments

The authors would like to thank Christophe Lamaze and Patricia Bassereau and all the people from the Membrane mechanics and dynamics of intracellular signaling laboratory and Membranes and Cellular Functions laboratory. The facilities as well as scientific and technical assistance from staff in the PICT-IBiSA/Nikon Imaging Centre at Institut Curie-CNRS and the France-BioImaging infrastructure (No. ANR-10-INSB-04) are acknowledged. This work was supported by institutional grants from the Curie Institute, INSERM, and CNRS, and by specific grants from Association Française contre les Myopathies (AFM): 22337 (to J.P.) and CAV-STRESS-MUS (14266 to C.M.B.). J.P. was funded by Polish Ministry of Science and Higher Education Mobility Plus program (1668/MOB/V/2017/0) and Labex CelTisPhyBio. The Lamaze team, the PICT-IBiSA/Nikon Imaging Centre at Institut Curie-CNRS, and the France-BioImaging infrastructure are members of Labex CelTisPhyBio (No. ANR-10-LBX-0038) and of IDEX PSL (No. ANR-10-IDEX-0001-02 PSL).

References

1. Martell JD, Deerinck TJ, Lam SS et al (2017) Electron microscopy using the genetically encoded APEX2 tag in cultured mammalian cells. Nat Protoc 12:1792–1816
2. Belkin M, Hardy WG (1961) Relation between water permeability and integrity of sulfhydryl groups in malignant and normal cells. J Biophys Biochem Cytol 9:733–745
3. Baumgart T, Hammond AT, Sengupta P et al (2007) Large-scale fluid/fluid phase separation of proteins and lipids in giant plasma membrane vesicles. Proc Natl Acad Sci U S A 104:3165–3170
4. Levental I, Byfield FJ, Chowdhury P et al (2009) Cholesterol-dependent phase separation in cell-derived giant plasma-membrane vesicles. Biochem J 424:163–167
5. Kaiser HJ, Lingwood D, Levental I et al (2009) Order of lipid phases in model and plasma membranes. Proc Natl Acad Sci U S A 106:16645–16650
6. Veatch SL, Cicuta P, Sengupta P et al (2008) Critical fluctuations in plasma membrane vesicles. ACS Chem Biol 3:287–293
7. Levental KR, Levental I (2015) Giant plasma membrane vesicles: models for understanding membrane organization. Curr Top Membr 75:25–57
8. Levental I, Lingwood D, Grzybek M et al (2010) Palmitoylation regulates raft affinity for the majority of integral raft proteins. Proc Natl Acad Sci U S A 107:22050–22054
9. Levental I, Grzybek M, Simons K (2009) Raft domains of variable properties and compositions in plasma membrane vesicles. Proc Natl Acad Sci U S A 108:11411–11416
10. Sezgin E, Kaiser HJ, Levental I (2012) Elucidating membrane structure and protein behaviour using giant plasma membrane vesicles. Nat Protoc 7:1042–1051
11. Parton RG, Hanzal-Bayer M, Hancock JF (2006) Biogenesis of caveolae: a structural model for caveolin-induced domain formation. J Cell Sci 119:787–796
12. Galbiati F, Razani B, Lisanti MP (2001) Caveolae and caveolin-3 in muscular dystrophy. Cell 106:403–411
13. Razani B, Lisanti MP (2001) Caveolins and caveolae: molecular and functional relationships. Exp Cell Res 271:36–44

14. Sinha B, Koster D, Ruez R et al (2011) Cells respond to mechanical stress by rapid disassembly of caveolae. Cell 144:402–413
15. Dewulf M, Koster DV, Sinha B et al (2019) Dystrophy-associated caveolin-3 mutations reveal that caveolae couple IL6/STAT3 signaling with mechanosensing in human muscle cells. Nat Commun 10:1974
16. Fridolfsson HN, Roth DM, Insel PA et al (2014) Regulation of intracellular signaling and function by caveolin. FASEB J 28:3823–3831
17. Lamaze C, Tardif N, Dewulf M et al (2017) The caveolae dress code: structure and signalling. Curr Opin Cell Biol 47:117–125
18. Collins BM, Davis MJ, Hancock JF et al (2012) Structure-based reassessment of the caveolin signaling model: do caveolae regulate signaling through caveolin-protein interactions? Dev Cell 23:11–20

Chapter 9

Biotin Proximity Labeling to Identify Protein–Protein Interactions for Cavin1

Carolina Mendoza-Topaz

Abstract

Biotin proximity labeling or BioID is a technique used to detect neighboring proteins, including transient and low-affinity interactions in their natural cellular environment. Here I describe the use of BioID in HeLa cells to identify proteins that can potentially interact with cavin1, one of the main components of caveolae. Briefly, the method consists in the transfection of the cells with the fusion constructs containing the promiscuous biotin ligase and cavin1 or control proteins, followed by biotin, cell lysis, affinity isolation of biotinylated proteins, biotin pull-down, and identification of the biotinylated proteins using mass spectrometry.

Key words Caveolae, Cavin1, caveolin1, Biotin proximity labeling, BioID, HeLa cells, Mass spectrometry, Biotin pull-down, Protein–protein interaction

1 Introduction

Current methods to identify interacting partners of caveolar proteins in cells are restricted to traditional coimmunoprecipation experiments, by using chemical crosslinking methods or in combination with cholesterol depletion or enrichment by subcellular fractionation experiments [1–3]. Coimmunoprecipitation experiments are potentially difficult to conduct and interpret due to the nature of Caveolin1 oligomers, as they are highly resistant to extraction with nonionic detergents [4, 5]. Approaches using chemically crosslinked samples have led to the characterization of a homogenous 80S complex formed by caveolin and cavin proteins at a specific stereochemistry (caveolar coat complex), but no further abundant component was found [3]. Nevertheless, caveolae have been associated with a surprisingly high number of proteins, in processes such as signaling, cell adhesion, and endocytosis [2]. The lack of evidence in many cases of a direct association (e.g., colocalization using high-resolution microscopy of those proteins with caveolin1 or immunoelectron microscopy) and the

Cedric M. Blouin (ed.), *Caveolae: Methods and Protocols*, Methods in Molecular Biology, vol. 2169,
https://doi.org/10.1007/978-1-0716-0732-9_9, © Springer Science+Business Media, LLC, part of Springer Nature 2020

questioning of the caveolin-signaling hypothesis [6] have led to the proposal that many of those interactions should be reassessed and skepticism should be maintained [1, 6].

Since caveolae are detergent-resistant structure, all classical immunoprecipitation experiments need to be carried out with a high concentration of detergent to fully solubilize caveolae. That may lead to disruption of potential authentic protein–protein interactions, thus making it extremely challenging to identify true transient and/or low-affinity interactions. Thus, to reassess or re-evaluate known interactions or search for potentially new partners, there is a need in the approach that is fundamentally different, and BioID appears as a very good candidate to overcome all of those challenges. Biotin proximity labeling [7] or BioID technique allows screening for interacting and neighboring proteins in their native cellular environment. The method relies on the fusion of a promiscuous biotin ligase to a targeted protein, and the biotinylated proteins are later isolated by affinity capture and identified by mass spectrometry. It must be noted, though, that not necessarily all the proteins identified by this technique will represent interacting partners to the protein of interest, but rather repertoire of proteins found in its close proximity.

In this chapter, I describe the use of biotin proximity labeling in cells to identify interaction partners of cavin1, one of the core components of caveolae [8]. The method is adapted from Roux et al. [7] and is based on the transient expression in HeLa cells of a fusion protein of cavin1 and the promiscuous biotin ligase BirA (R188G) (BirA*). The method requires the addition of biotin for a long period of time (16–24 h) so that any protein found in close proximity to cavin1-BirA* protein (approx. radius of 20 nm) would be biotinylated at any time. Recent engineered versions of promiscuous mutants of biotin ligase, such as TurboID [9], can work in the range of minutes. TurboID can potentially be used in the future to identify caveolae-associated proteins in a more restricted window time or even comparing the biotinylated protein repertoire with or without certain stimulus done for a short period of time (like the stretching of the plasma membrane).

The controls used here are a set of other membrane or membrane-associated proteins, similarly fused with BirA* that induces biotinylation of plasma membrane proteins. The controls used are: BirA* with a myristoylation and palmitoylation sequence (MyrPalm-BirA*), a combination of flotillins 1 and 2 (Flot1/2-BirA*), and CD20-BirA*. The method allowed the identification of known caveolar components (cavin1, cavin3, and caveolin1) as well as new ones (CSDE1, Vigilin, CKAP5, etc.) (*see* **Note 1**).

The protocols and methods described here consist of the following steps:

(a) Generation of myc-BirA* DNA constructs: Cavin1, MyrPalm, Flot1/Flot2, and CD20.
(b) Transformation of constructs into cells (day 1) (*see* **Note 2**).
(c) Addition of biotin (day 2).
(d) Washes and cell lysis (day 3) (*see* **Note 3**).
(e) Biotin pull-down with magnetic beads (day 4).
(f) Bead washes and elution (day 5).
(g) Experimental quality assurance by anti-caveolin1 and anti-myc Western blots analysis (to test whether caveolin1 is being pulled-down in cavin1-BirA* samples and to test loading control, respectively) (day 6).
(h) Run the rest of the samples in SDS acrylamide gel and cut out gel pieces for mass spectrometry (day 7).
(i) Mass spectrometry.
(j) Repeat **steps (b)** to **(i)** three times.
(k) Data analysis.

2 Materials

2.1 DNA Constructs and General DNA Manipulation

1. Backbone vector: mycBirA*(R118G) in pCDNA3.1 [8, 10].
2. Inserts: full-length rat cavin1, mouse flotillin1, rat flotillin2, human CD20 (*see* **Note 4**), and a myristoylation and palmitoylation sequence (GCGCSSHPEDDGGSGGSGGS) (*see* **Note 5**).
3. DNA manipulation reagents:
   (a) Agarose gel, DNA molecular weight marker, SYBR safe, electrophoresis chambers, power pack.
   (b) NheI and BamHI restriction enzymes, Gibson assembly master mix, 37 and 50 °C heating block, gel extraction kit, 1.5-mL tubes, scalpel, UV transilluminator.
   (c) T4 DNA ligase and buffer, 100% and 70% ETOH, 3 M NaAcetate of pH 5.2, chemically competent bacteria for transformation.

2.2 Tissue Culture

1. Low-passage HeLa cells (<15).
2. DMEM medium supplemented with penicillin/streptomycin and 10% of heat-inactivated fetal bovine serum (FBS).
3. 37 °C, 5% CO_2 incubator, 1× trypsin-EDTA solution. Tissue culture dishes (10 and 15 cm) (*see* **Note 6**).

2.3 Transformation

1. FugeneHD or PEIMAX (Polyethyleneimine MW 40,000).
2. Opti-MEM.
3. DMEM medium supplemented with 10% FBS without antibiotics.

2.4 Biotin Addition

1. DMSO.
2. Biotin powder.

2.5 Cell Lysis

1. PBS, 2-mL tubes, ice.
2. Rubber or silicone cell scrapers with pivoting blades.
3. 40% *n*-octylglucoside (OG) stock (stock aliquoted and kept at −20 °C) (*see* **Note 7**).
4. TritonX100.
5. EDTA-free protease inhibitors cocktail.
6. Lysis buffer: 50 mM Tris–HCl of pH 8, 300 mM NaCl, 5 mM EDTA, 1% Triton X-100, 1% *n*-octylglucoside (OG), and complete protease inhibitor cocktail. Prepare 50 mL without OG and keep it at 4 °C. Add 250 μL of 40% OG per 10-mL lysis buffer just before use.

2.6 Biotin Pull-Down

1. MyOne Streptavidin C1 beads.
2. Bead washing buffer (1 lysis buffer : 1 50 mM Tris–HCl of pH 7.4).
3. Washing solutions as described in Roux et al. [7]:
   (a) Wash buffer 1: 2% SDS.
   (b) Wash buffer 2: 1% Triton X-100, 0.1% deoxycholate, 500 mM NaCl, 1 mM EDTA, 50 mM HEPES.
   (c) Wash buffer 3: 250 mM LiCl, 0.5% NP-40 alternative (IGEPAL), 0.5% deoxycholate, 1 mM EDTA, 10 mM Tris–HCl of pH 8.
   (d) Wash buffer 4: 0 mM Tris–HCl of pH 7.4, 50 mM NaCl. All solutions with protease inhibitors.
4. Magnetic stand for 1.5–2-mL tubes.
5. 2-mL tubes.
6. Rotator for 2-mL tubes (to be used at 4 °C).
7. 1–2.5 M DTT stock (stocks aliquoted and kept at −20 °C).
8. 4× LDS sample buffer.

2.7 Protein Analysis by Western Blot

1. 4–12% Bis-Tris SDS-PAGE gels, 1-mm thick, 15 wells.
2. Power pack.
3. Prestained protein molecular weight marker.

4. 1× MES SDS running buffer. Prepare 20× stock as follow: 97.6 g MES, 60.6 g Tris–HCl base, 10 g SDS, and 3 g EDTA dissolved in 400 mL of MPW, mix well, and make final volume to 500 mL. Store at 4 °C covered in foil to protect from light.
5. Whatman filter paper, PVDF membranes, Methanol, transfer tank for wet transfer.
6. Transfer buffer. Prepare 20× stock (without methanol) as follow: 30 g Tris–HCl base and 144 g glycine in 1 L. For 2 L 1× transfer buffer, mix 200 mL 20× Tris–HCl/glycine stock, 400 mL Methanol, and make final volume to 2 L.
7. 0.1% Tween-20 in 1× PBS (PBST solution).
8. 5% dried skimmed milk powder in PBST.
9. Shaker for blots.
10. Streptavidin-HRP.
11. Antibodies: rabbit anti-Cav1 (BD Biosciences 610060), rabbit anti-Cavin1 (Abcam ab48824), mouse anti-myc, anti-mouse-HRP, anti-rabbit-HRP.
12. iBlock (or other protein-based blocking reagent).
13. Chemiluminescent Western Blot Detection Reagent Kit
14. Cassettes for X-ray films and X-ray films.

2.8 Bands for Mass Spectrometry

1. 4–12% Bis-Tris SDS-PAGE gels, 1.5-mm thick, 10 wells.
2. Colloidal Coomassie.
3. Scalpels (at least 50).
4. Methanol-cleaned glass (that has never been treated with detergents).
5. Clean microtiter 96-well plate with lid.
6. HPLC-grade water.

2.9 Immuno-fluorescence (Optional)

1. Sterile 22 mm × 22 mm coverslips.
2. 6-well plates.
3. 4% PFA.
4. Parafilm.
5. Coverslip tweezers.
6. Plastic container (to be used as humid chamber for antibody staining).
7. Blocking solution: 0.2% saponin, 10% FBS in PBS.
8. Fluorescent conjugate antibodies: Streptavidin-647 and Alexa Fluor fluorescent secondary antibodies (488, 555).
9. Glass slides for imaging.
10. Mounting medium.

2.10 Mass Spectrometry

Consult with the specialist who will run the samples.

2.11 Analysis Data

1. Scaffold program.
2. Statistical analysis.

3 Methods

Carry out all the DNA and protein procedures in a laboratory bench at room temperature unless otherwise specified and tissue culture work in a tissue culture room inside a tissue culture hood with septic technique. Cells are incubated in a 37 °C incubator with 5% CO_2. All centrifugations are carried out at room temperature unless otherwise specified.

3.1 BirA* Constructs

1. Inserts (Cavin1, flotillin1, flotillin2, CD20, and myristoylation and palmitoylation sequence (MyrPalm)) were cloned by traditional molecular biology methods into a pcDNA3.1(+) plasmid containing BirA*(R118G) [10] followed by a GSGSGS linker and a myc tag (EQKLISEEDL) [7].
2. Purify plasmids using midiprep kit (*see* **Note 8**). Perform ethanol precipitation of the DNA if the yield is lower than 500 ng/mL and resuspend DNA pellet in 10 mM Tris–HCl, pH 8.5, to bring to final DNA concentration 1–1.5 μg/mL.

3.2 Expansion of Cells

HeLa cells for first round of transfections.

1. Day 1: Thaw a low passage 2-mL cryotube (approx. five million cells) of HeLa cells at 37 °C and transfer them to one 15-cm tissue culture dish with 20 mL of prewarmed (37 °C) DMEM with 10% heat-inactivated FBS with pen/strep. Leave overnight at 37 °C in the incubator.
2. Day 2: the cells should be approximately 80% confluence. If cells are not ready (less than 70% confluence), change medium and wait an additional day. If cells are confluent enough, aspirate medium, wash cells with PBS, aspirate PBS, and add 2 mL trypsin. Incubate 5 min at 37 °C.
3. In the meantime, prepare three 15-cm dishes with 16 mL of prewarmed medium each one.
4. Remove cells from incubator and pipette cells up and down.
5. Add approximately 660 mL of cells to each dish (1/3 of original). Mix well and incubate overnight (*see* **Note 9**).
6. Day 3 (or day 4 when cells are confluent). Repeat **step 2**.
7. In the meantime, prepare ten 15-cm dishes with only 12 mL of prewarmed medium without antibiotics in each dish.

8. Repeat **step 4**.
9. Pool all the cells (6 mL total) and add them to a 50-mL Falcon tube with 24-mL prewarmed medium. Mix well.
10. Spin cells at 250 × *g* for 5 min. Remove medium and add 48 mL prewarmed medium without antibiotics. Mix well.
11. Add 4-mL cells to each dish with the 16-mL medium already added, mix gently, and incubate overnight at 37 °C. Discard the rest of the cells or keep them for expansion.

3.3 Transfection

For efficient transfection, cells should be around 70% confluence (*see* **Note 10**).

1. Inside tissue culture hood, prepare 4× 1.5 mL tubes with the plasmid DNA: tube Cavin1 with 10 μg Cavin1-mycBirA*, tube Flots with 5 μg Flot1-mycBirA* and 5 μg Flot2-mycBirA*, tube MyrPalm with 10 μg MyrPalm-mycBirA*, and tube CD20 with 10 μg CD20-mycBirA*. Add 500 μL Opti-MEM per tube. Incubate at room temperature for 5 min.
2. Prepare a 2-mL tube with FugeneHD or your preferred transfection reagent with Opti-MEM. For a ratio 1:5 (1 μg DNA : 5 μL FuGeneHD), mix 200 μL FugeneHD with 1800 μL of Opti-MEM. Incubate at room temperature for 5 min.
3. Add 500 μL of DNA-liposome complexes to each tube with the DNA, vortex, and incubate at room temperature for 10 min.
4. In the meantime, label the 10 × 15 cm dishes as follows: 2 for Cavin1, 2 for Flot1/Flot2, 2 for MyrPalm, 2 CD20 and 2 as nontransfected.
5. Add 500 μl of the complexes to each dish accordingly. Incubate overnight at 37 °C.

3.4 Addition of Biotin

1. Next day, just before changing medium to the cells, prepare 50 mM biotin stock by adding 12 mg Biotin to 1 mL DMSO in a 1.5-mL tube.
2. Take transformed cells from incubator and aspirate the medium.
3. Add 200 μL of the freshly made 50 mM biotin to 200-mL medium without antibiotics, mix well, and add 20 mL to each of the 10 × 15 cm dishes, including to the nontransformed ones (negative control). Alternatively, add 200 μL of the freshly made 50 mM biotin to 50-mL medium, mix well, and add 5 mL to each dish with 15-mL fresh medium, no antibiotics. Mix well. Incubate overnight at 37 °C.

3.5 Cell Lysis

1. After 20 h of biotin addition (*see* **Note 11**), prepare fresh lysis buffer.
2. Take cells from the incubator and wash them thoroughly six times with PBS.
3. Remove thoroughly all PBS and quickly proceed to the cell lysis step.
4. Add 1-mL lysis buffer per dish to all the dishes, meticulously scrap cells with a cell scraper, and transfer lysates to 1.5-mL tubes.
5. Vortex briefly and incubate on ice for 30 min.
6. Vortex briefly and spin at 16K × *g* for 15 min at 4 °C.
7. Transfer supernatant to a 2-mL tube and add 1 mL of 50 mM Tris–HCl of pH 7.4. Mix by inversion.
8. Save 100 μL for Western blot analysis.

3.6 Biotin Pull-Down

1. Label 10 × 2 mL tubes, add 500 μL of Dynabeads MyOne Streptavidin C1 beads in each tube (*see* **Note 12**), and add 1-mL bead washing buffer (1 lysis buffer : 1 50 mM Tris–HCl of pH 7.4).
2. Stand tubes in the magnetic stand for 2 min (to allow all the beads to be tightly bound to the magnet), aspirate, and repeat washing.
3. Aspirate washing buffer from beads and add lysates. Incubate overnight at 4 °C with rotation.

3.7 Bead Washes and Elution

1. Next day, stand tubes in magnetic stand for 2 min, save 100 μL supernatant (unbound fraction), aspirate, and discard the rest of the supernatant.
2. Wash 2× for 8 min with 1-mL wash buffer 1 with rotation at room temperature leaving the tubes on the stand for 2 min between washes.
3. Continue washing 2 × 8 min with buffer 2, once ×8 min with buffer 3 and 2 × 8 min with buffer 4.
4. During the washes, prepare fresh biotin solution by adding 15-mg biotin to 10-mL distillate water (*see* **Note 13**). Heat at 80 °C, mix, and use immediately. Do not store.
5. Prepare elution buffer containing 1.5× LDS elution buffer, 150 mM DTT, and half of the volume with the biotin solution. Heat at 95 °C.
6. Remove from the beads as much as possible of the last wash and add 30 μL of hot elution buffer.
7. Incubate at 98 °C for 5 min, stand tubes in magnetic stand, collect eluates, and keep them.

8. Add another 30 μL of hot elution buffer to each tube and repeat **step 7**.
9. Combine eluates of each sample (60 μL final volume per sample). Keep samples at −20 °C or run them immediately for a Western blot analysis or use directly for mass spectrometry.

3.8 Western Blot Test (See Note 14)

1. Run 5 μL of each duplicate sample in 4–12 Bis-Tris SDS-PAGE gels, including untransfected samples and prestained protein marker. Another gel with the same samples can be run in parallel to test the pattern of biotinylation (*see* **Note 15**).
2. Soak a PVDF membrane in methanol and use it for a wet transfer for 4 h × 40 V at 4 °C or overnight × 25 V at 4 °C. Take membranes from the transfer tank, soak them in methanol, and let them completely dry at room temperature.
3. Incubate with rabbit anti-cav1 antibody 1:10,000 (*see* **Note 16**) in 5% milk PBST for 2 h at room temperature or overnight at 4 °C, wash 5 × 5 min with PBST, and then incubate 1 h at room temperature with anti-rabbit HRP secondary antibody.
4. If another duplicated gel was run in parallel to test biotinylation pattern, block with 5% milk in PBST or with 0.2% iBlock × 1 h room temperature (*see* **Note 17**) and then incubate for 1 h at room temperature with 1:4000 Streptavidin-HRP.
5. Wash 5 × 5 min with PBST and develop with chemiluminescent reagent (*see* **Note 18**).
6. Reprobe anti-cav1 PVDF membrane with anti-myc antibody (*see* **Note 19**).

3.9 Immunofluorescence Test (Optional) (See Note 20)

1. Place one sterile 2 mm × 22 mm coverslip inside each well of a 6-well plate. Seed cells on top, so they are 70% confluent the next day (in 1.5-mL medium).
2. Next day transfect cells (as on Subheading 3.3) but using 500 ng DNA for each well (instead of the 5 μg used for each 15-cm dish) in 100 μL Opti-MEM and 2.5 μL FuGeneHD in 100 μL Opti-MEM. Transfect one construct per well and leave one well with cells as mock control.
3. Next day, change medium with biotin (as on Subheading 3.4) and incubate for 20 h at 37 °C.
4. Wash cells thoroughly six times with PBS and fix cells with 4% PFA for 15 min at room temperature and wash with PBS.
5. Block and permeabilize cells with 0.2% saponin, 10% FBS in PBS for 30 min at room temperature.
6. Prepare a humid chamber by placing flat and tight a piece of Parafilm long enough for the 6 coverslips inside a plastic container with a lid. Soak with water two pieces of paper and leave them on two opposite edges of the container without touching the Parafilm.

7. Prepare primary antibody mix as follows: Rabbit anti-caveolin1 (1:3000) and mouse anti-myc (dilution depending on the antibody used) in blocking solution. Add 80–100 μL drops of antibody mix on top of the Parafilm of the humid chamber.
8. Take coverslips carefully using coverslip tweezers, remove the excess of liquid by touching a tissue paper, and place them on the drop of antibody, with the cells facing down. Incubate for 1 h at room temperature or overnight at 4 °C.
9. Take coverslips carefully using coverslip tweezers and place them back on the 6-well plate with the cells facing up. Wash 4× with PBS.
10. Prepare secondary antibody mix as follows: anti-rabbit Alexa 555, anti-mouse Alexa 488, and Streptavidin-647 (all 1:500). Take coverslips as in 10 and incubate 1 h at room temperature with the secondary antibody mix.
11. Repeat 9.
12. Mount coverslips on glass slides for imaging (Fig. 1).

3.10 Samples for Mass Spectrometry

Gloves should be worn at all times and be as clean as possible at all stages of sample preparation starting from cell lysis, gel running, and subsequent manipulation. All material (plastic trays for staining, glass to cut bands, etc.) should never be with washed or in contact with detergents. Avoid preparing samples directly under air-conditioning units. Before running the samples in the gel and cutting the bands to be sent for mass spectrometry, consult first with the specialist who will run the samples for mass spectrometry. Details of mass spectrometry technique are beyond the scope of this chapter and are not described here.

1. Run 50 μL of each eluted sample in a 4–12% Bis-Tris SDS-PAGE gel, 1.5-mm-thick or 1-mm gel suitable for 50 μL. Run for 4–5 cm.
2. Stain gel with Colloidal Coomassie InstantBlue for 1 h. Discard and replace with water.
3. Carefully place the gel on a methanol-washed glass. Observe on a well-lit hood (Fig. 2).
4. The excision of bands should be done with surgical precision. Cut the entire gel lanes into eight slices per lane and place each band in a well of an appropriate 96-well plate with 20–50 μL HPLC water. Gel bands should be between 1 and 2 mm thick, with a minimal area of nonstained gel.
5. Samples can be sent directly for mass spectrometry analysis or kept overnight at 4 °C.
6. MS/MS data are validated using the Scaffold program (Proteome Software Inc).

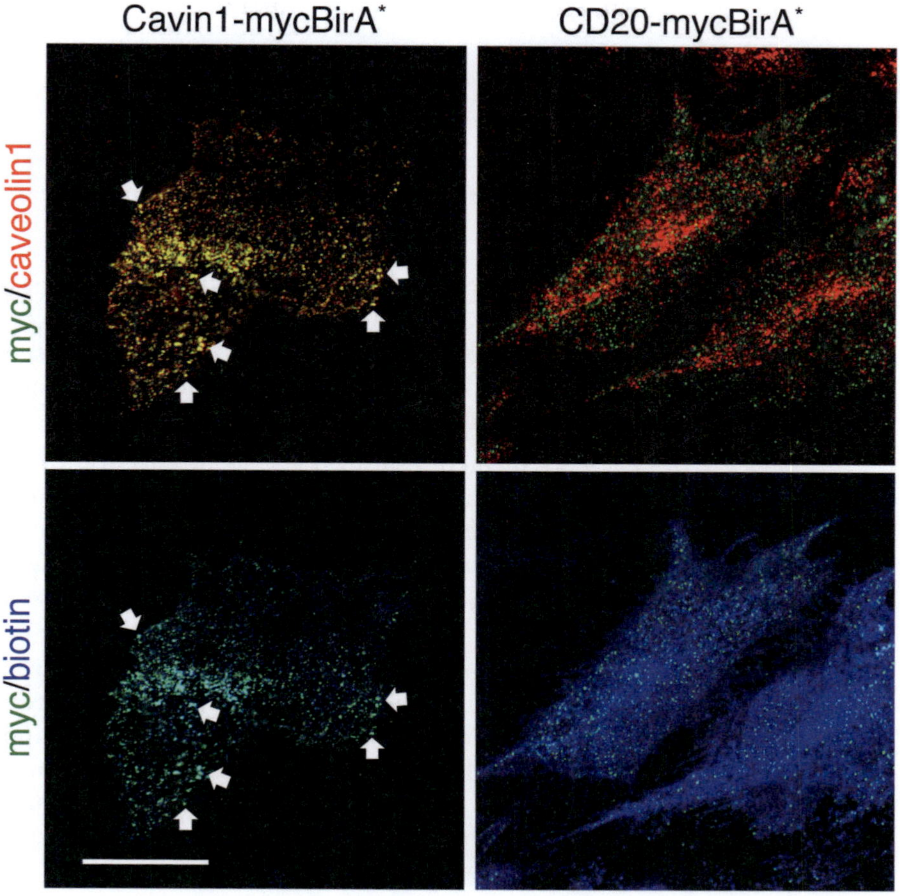

Fig. 1 Cells are transfected with cavin1-myc-BirA* (left panel) or with CD20-mycBirA* (right panel) and stained with anti-myc (green) and anti-caveolin1 (cav1) (red) or biotin (blue). Only cavin1-myc-BirA* colocalizes with cav1 (yellow dots on the top-left panel), gets biotinylated (green and blue dots on the bottom-left panel), and induces biotinylation of caveolae (white arrows show colocalization of cav1 staining and biotin). Bar is 30 μm

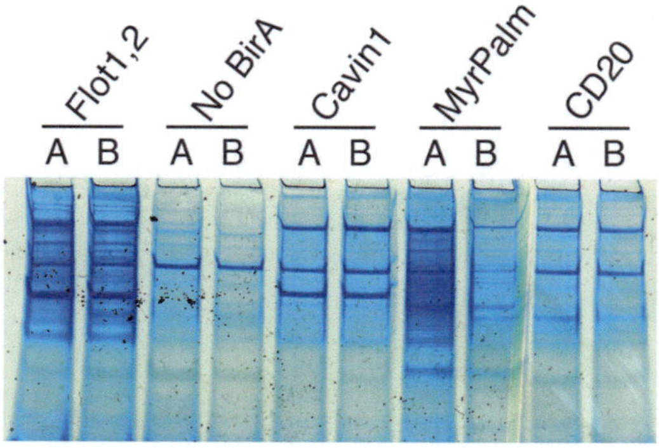

Fig. 2 Example of a stained gel before the bands are cut. A and B are two biological replicates

3.11 Experimental Replicates and Analysis

1. Repeat steps in Subheadings 3.2–3.7 and 3.10 two more times to have three independent experiments with two biological replicates each.
2. Perform statistical analysis (*see* **Note 21**).

4 Notes

1. Once a new potential interacting protein is identified, the researcher can evaluate the possibility to do the reverse experiment if adequate controls can be established. For example, test if the putative cavin1- or caveolin1-interacting protein fused with BirA* is able to biotinylate caveolae.
2. The current method uses transient transfection of the BirA* constructs. Before biotin pull-down, it is recommended to confirm the expression level of overexpressed protein in protein lysates by Western-blot analysis with anti-myc antibodies. Alternatively, stable cell lines with an inducible system can be used.
3. Cavin1 contain nuclear localization sequences and may have nuclear functions independent of caveolae. A researcher interested in the nuclear functions of cavin1 can design appropriate nuclear controls, perform a subcellular fractionation after cell lysis and focus only on the nuclear fraction. The nuclear fraction can be eliminated if the researcher wants to exclude Cavin1 nuclear functions.
4. Some BirA* constructs suggested here use cDNA from other species not human; these were the constructs used in Mendoza-Topaz et al. [8]. Rat cavin1 has 94% identity to human cavin1. Mass spectrometry can discriminate between exogenous and endogenous proteins based on the species, for example, rat cavin1 biotinylates itself and endogenous human cavin1 and cav1. At the same time, mouse and rat flotillins can induce biotinylation of human flotillins in HeLa cells. This analysis is useful for internal control to assure that overexpressed exogenous proteins were properly localized and assembled with their target partners.
5. All of these controls would induce biotinylation of the plasma membrane. The researcher can choose other or a complementary set of plasma membrane/plasma membrane–associated proteins.
6. Cell lysis can be done easily by directly scraping cells in tissue culture dishes instead of growing the cells in tissue culture flasks, followed by trypsinization, centrifugation, and then the washes of the cell pellets in tubes.

7. *N*-octylglucoside should be added fresh to the lysis buffer just before use.
8. Plasmid DNA should be at a concentration higher than 500 ng/mL, so no more than 10 μL is used per transfection (5 μg DNA per 15-cm dish). DNA concentration and quality (ratio of *A*260/*A*280 should be 1.8–1.9) can affect the efficiency of transfection.
9. For simplicity, I am going to refer to the amount of cells used in each step as a proportion of the confluent dish, but if the cells chosen behave differently, the amount of cells needs to be adjusted accordingly. For transfections, seed the amount of cells necessary to have cells at around 70% confluence next day.
10. Hela cells transfect very efficiently with different transfection reagents. FuGeneHD has an efficiency of 80–90% but is very sensitive to cell confluency. In our hands, if cells were below 60% confluency, the efficiency of transfection with FuGeneHD was very low and cells start to die soon after. If cells are too confluent (more than 80%), they do not transform very well. PEI is more versatile and allows a wide range of confluences with reasonable good efficiency of transfection but seems to be more toxic.
11. Once biotin is added, the promiscuous biotin ligase would biotinylate all the proteins encountered in its close proximity from when the protein is translated. Therefore, the longer incubation with biotin, potentially more proteins would get biotinylated (and hence more background). We found that with only 8 h of biotin incubation, cavin1-mycBir*A is able to biotinylate caveolin1 when biotinylated proteins are pulled-down and caveolin1 is analyzed by Western blot (Subheading 3.8). However, caveolin1 was never found in mass spec data from the 8-h point. Caveolin proteins are difficult to detect by mass spectrometry, probably because of their small size and few suitable peptides generated after trypsin digestion for mass spectrometry and also presumably because of their high hydrophobicity [3, 8]. The only cavin proteins found in caveolar complexes in HeLa cells are cavin1 and cavin3. Cavin2 is not expressed in HeLa cells [3, 8]. With 16–24 h after biotin addition, all the components of caveolar complex in HeLa cells can be found significantly enriched by mass spectrometry and by Western blotting (caveolin1, cavin1, and cavin3). Caveolin2 was never found with the exception of few experiments where it was found only in the cavin samples.
12. Unfortunately, it is necessary to use this large amount of beads. With 200-μL beads per sample, we did not detect caveolin1 with cavin1-mycBirA* in the mass spec data (even though it is detected by Western blotting).

13. Solubility of biotin in H_2O is 0.2 mg/mL (0.82 mM); therefore, preparing a solution 15 mg/mL (6 mM) is oversaturated. Some white powder will remain in solution but that will not affect the elution. That is the reason for the repeated elution step. If the beads are re-eluted for a third time, run in a gel, and blotted for caveolin1, it is still possible to detect some caveolin1 in cavin1-mycBirA* samples. That can be due to a combination of factors, such as the difficulty to completely deplete the beads and the high quality of the caveolin1 antibody.
14. Caveolin1 is only found in Cavin1-mycBirA* samples when biotin is added and not in the controls. Very long exposure times in Western blots can detect some caveolin1 in MyrPalm-BirA* samples, but not Flotillins or CD20. Caveolin1 was never found significantly enriched in the mass spec data of any of the controls.
15. This can be done only the first time the experiment is done to check (1) That many discrete proteins get biotinylated with each BirA* construct. (2) The presence of an intense biotinylation band of the same size as each construct used (self-biotinylation of constructs). Sizes of myc-BirA* constructs are: cavin 88 kDa, Flot1/Flot2 80 kDa, CD20 71 kDa, and MyrPalm 39 kDa. (3) Similarity of the biological duplicates.
16. Anti-cav1 antibody can be recycled many times. Can be kept at 4 °C in PBST with sodium azide at a final concentration of 0.02% for over 6 months or in PBST with 5% milk at −20 °C.
17. iBlock or similar gives less background as it is a biotin-free blocking reagent.
18. It is not recommended to use supersensitive reagents for anti-caveolin1 Western blot even for 1:10,000 dilution, as the signal will be oversaturated and limited to a very short period of time.
19. In case efficiency of transfection varies greatly between samples, normalize the amount to be loaded for mass spectrometry gel-based on anti-myc blot from inputs.
20. This test can be done when the researcher uses the constructs for the first time. The staining can be done on coverslips (as described here) or in m-slide 8 or 4-well chambers (Ibidi) or similar. Those chambers can be used directly for confocal imaging, so there is no need for transferring the cells from one compartment to another for washes and incubations with antibodies.
21. In Mendoza-Topaz [8], an enrichment score was calculated for each protein in each experiment, based on the exclusive spectrum counts of cavin1 relative to all the other controls in that experiment. Then a normalized product of scores was generated by multiplying together (enrichment score +1) for each

experiment and then normalizing to the result in the case of caveolin1. Other way to analyze the data is using a *D*-score from CompPASS platform, derived from *S*-score [11]. Other parameters that can be added to the analysis are to consider the size of the protein and/or the abundance of the proteins in the cell type analyzed [12].

References

1. Cheng JPX, Nichols BJ (2016) Caveolae: one function or many? Trends Cell Biol 26:177–189
2. Parton R, Simons K (2007) The multiple faces of caveolae. Nat Rev Mol Cell Biol 8:185–194
3. Ludwig A, Howard G, Mendoza-Topaz C, Deerinck T, Mackey M, Sandin S, Ellisman MH, Nichols BJ (2013) Molecular composition and ultrastructure of the caveolar coat complex. PLoS Biol 11:e1001640
4. Kovtun O, Tillu VA, Ariotti N, Parton RG, Collins BM (2015) Cavin family proteins and the assembly of caveolae. J Cell Sci 128:1269–1278
5. Hayer A, Stoeber M, Bissig C, Helenius A (2009) Biogenesis of caveolae: stepwise assembly of large caveolin and cavin complexes. Traffic 11:361–382
6. Collins BM, Davis MJ, Hancock JF, Parton RG (2012) Structure-based reassessment of the caveolin signaling model: do caveolae regulate signaling through caveolin-protein interactions? Dev. Cell 23:11–20
7. Roux KJ, Kim DI, Raida M, Burke B (2012) A promiscuous biotin ligase fusion protein identifies proximal and interacting proteins in mammalian cells. J Cell Biol 196:801–810
8. Mendoza-Topaz C, Yeow I, Riento K, Nichols BJ (2018) BioID identifies proteins involved in the cell biology of caveolae. PLoS One 13: e0209856
9. Branon TC, Bosch JA, Sanchez AD, Udeshi ND, Svinkina T, Carr SA, Feldman JL, Perrimon N, Ting AY (2018) Efficient proximity labeling in living cells and organisms with TurboID. Nat Biotechnol 36:880–887
10. Shin JJH, Gillingham AK, Begum F, Chadwick J, Munro S (2017) TBC1D23 is a bridging factor for endosomal vesicle capture by golgins at the *trans*-Golgi. Nat Cell Biol 19:1424–1432
11. Sowa ME, Bennett EJ, Gygi SP, Harper JW (2009) Defining the human deubiquitinating enzyme interaction landscape. Cell 138:389–403
12. Nagaraj N, Wisniewski JR, Geiger T, Cox J, Kircher M, Kelso J, Pääbo S, Manna M (2011) Deep proteome and transcriptome mapping of a human cancer cell line. Mol Syst Biol 7:548

Chapter 10

Investigation of Novel Cavin-1/Suppressor of Cytokine Signaling 3 (SOCS3) Interactions by Coimmunoprecipitation, Peptide Pull-Down, and Peptide Array Overlay Approaches

Jamie J. L. Williams, George S. Baillie, and Timothy M. Palmer

Abstract

The ability of inducible regulator suppressor of cytokine signaling 3 (SOCS3) to inhibit Janus kinase-signal transducer and activator of transcription (JAK-STAT) signaling requires interaction with specific cytokine receptors, JAKs, and components of the cellular ubiquitylation machinery. However, it is now clear that additional protein interactions are essential for effective inhibition of JAK-STAT signaling that have also identified new roles for SOCS3. For example, we have demonstrated that SOCS3 interaction with cavin-1, a core component of caveolae essential for their formation, is required for effective inhibition of interleukin (IL)-6 signaling and maintenance of cellular levels of caveolae. This is achieved through cavin-1 interaction with a discrete motif within the SOCS3 SH2 domain. Here, we describe in detail three methods (coimmunoprecipitation; peptide pull-down; peptide array overlay) we have used to validate and characterize cavin-1/SOCS3 interactions in vitro.

Key words Coimmunoprecipitation, Peptide array, Peptide pull-down, Protein-protein interactions, JAK-STAT, Caveolae, Cavin-1, SOCS3

1 Introduction

Suppressor of cytokine signaling 3 (SOCS3) is one member of a family of eight (cytokine-inducible SH2-containing protein (CIS), SOCS1–7) inducible, negative feedback inhibitors of cytokine receptor signaling [1]. SOCS3 terminates signaling from gp130, a ubiquitously expressed membrane/lipid raft-localized signaling receptor required for downstream signaling by IL-6 family cytokines. Ligand-activated gp130 is transphosphorylated on specific tyrosine residues by constitutively bound Janus kinases (JAKs), which triggers the recruitment of signal transducers and activators of transcription (STATs). After phosphorylation by receptor-bound JAKs, STATs dissociate from the receptor, dimerize, and translocate

Cedric M. Blouin (ed.), *Caveolae: Methods and Protocols*, Methods in Molecular Biology, vol. 2169,
https://doi.org/10.1007/978-1-0716-0732-9_10,

to the nucleus to drive transcription of several genes, including SOCS3 [2]. SOCS3 binds PTyr759 on gp130 *via* a central SH2 domain. This enables it to inhibit JAKs *via* its N-terminal kinase inhibitory region (KIR), thereby preventing binding and activation of STATs [3]. In its role as a specificity factor within an E3 ubiquitin ligase complex, SOCS3 also targets JAKs for proteasomal degradation [4]. SOCS3 binds elongins B and C, Cullin family member Cul5, and RING (Really Interesting New Gene) finger protein Rbx2 *via* its C-terminal SOCS box domain, with target substrates binding *via* the central SH2 domain [3] (Fig. 1a).

A novel SOCS3-interacting protein, CUE domain-containing 2 (CUEDC2), has been shown to enhance its interaction with elongin C, raising the possibility that other SOCS3 interactors might modulate SOCS3 function [5]. We have recently identified cavin-1 (also known as polymerase I and transcript release factor (PTRF)), an abundant component of caveolae, as a novel SOCS3-interacting protein [6]. Cavin-1 and related family members (cavin-2–4) are recruited to the plasma membrane by caveolin proteins (caveolin-1–3) to form and stabilize caveolae [7]. We found that SOCS3 directly binds cavin-1 through an interaction that requires the SH2 domain-localized PEST (Pro-Glu-Ser-Thr) motif. Furthermore, cavin-1 was necessary for SOCS3-mediated inhibition of IL-6 signaling, while the presence of SOCS3 stabilized cavin-1 and thus maintained levels of cell surface caveolae. Therefore, our findings define a novel link between SOCS3 and cavin-1 in which each partner plays previously unappreciated roles in maintaining effective inhibition of JAK–STAT signaling (cavin-1) and cell surface caveola through stabilization of cavin-1 (SOCS3) [6].

In this chapter, we provide three protocols to investigate SOCS3/cavin-1 interactions. Coimmunoprecipitation (co-IP) is the precipitation, *via* centrifugation, of interacting proteins following the formation of an immune complex bound to a support bead. Peptide pull-down is a technique similar to co-IP but, in place of an antibody, a short peptide from a protein of interest immobilized to a support bead is used as bait to precipitate protein complexes. In each case, interactions are analyzed by immunoblotting. Peptide array overlay allows direct interactions to be investigated while also allowing the identification and validation of residues or domains responsible for the interaction. In this protocol, a full polypeptide sequence is spotted onto a support membrane as 25-mer peptides that overlap by five residues. The membrane is then overlaid with either purified protein or soluble lysates from cells transiently overexpressing the protein of interest and interactions on the array detected using a primary antibody against the protein target. All methods allow visualization by standard gel documentation systems.

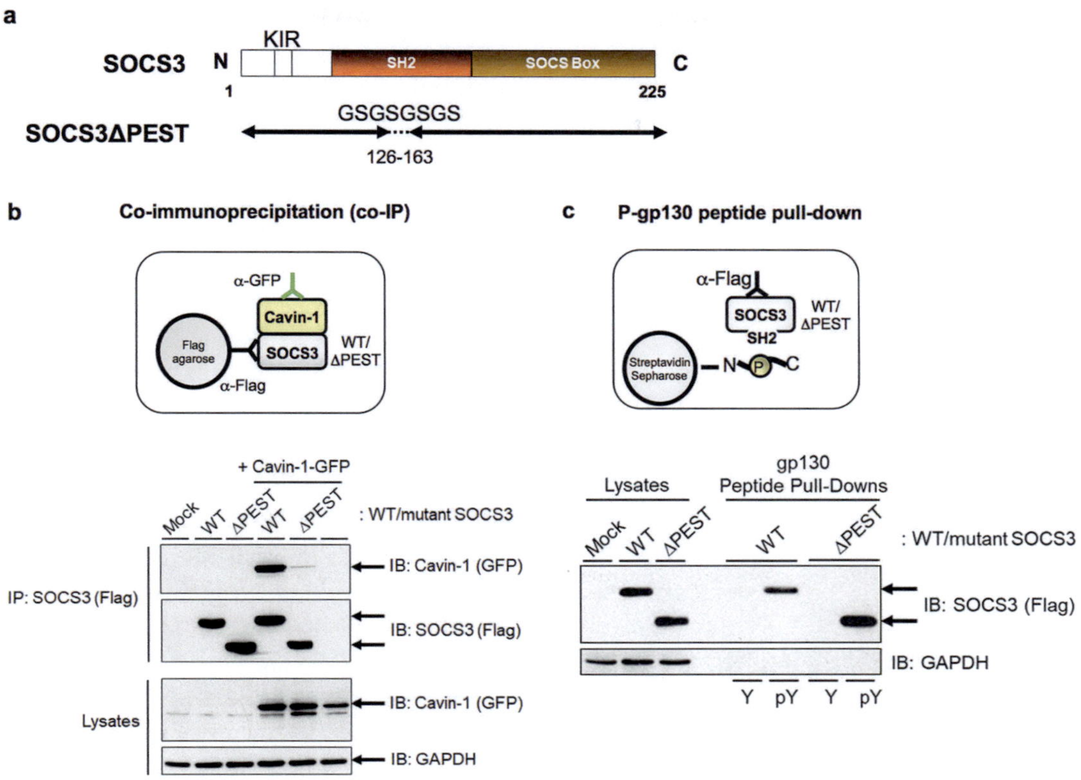

Fig. 1 Investigation of SOCS3/cavin-1 interaction *via* coimmunoprecipitation and peptide pull-down. (**a**) Domain structure of SOCS3 and the SOCS3ΔPEST mutant in which the PEST motif is replaced with a Gly-Ser (GS) repeat. (**b**) Protein-equalized soluble cell extracts from HEK293 cells transfected with expression constructs encoding either wild-type (WT) or ΔPEST Flag-SOCS3 and cavin-1-GFP as indicated were processed by co-IP with anti-Flag M2 antibody-agarose beads prior to SDS-PAGE and immunoblotting (IB) with the indicated antibodies. Whole-cell lysates from the samples used in the co-IP were also immunoblotted in parallel. (**c**) Protein-equalized soluble cell extracts from HEK293 cells transfected with expression constructs encoding either WT or ΔPEST Flag-SOCS3 were incubated with 100 nM N-terminally biotinylated peptides corresponding to the Tyr759 motif of gp130 in its phosphorylated (pY) or nonphosphorylated (Y) form and streptavidin–agarose beads as indicated in the schematic diagram. Peptide pull-down samples and whole-cell lysates were then analyzed by SDS-PAGE and immunoblotting (IB) with the indicated antibodies

2 Materials

2.1 Coimmunoprecipitation

1. Human embryonic kidney (HEK) 293 cells maintained in Dulbecco's modified Eagle's medium (DMEM) supplemented with 10% (v/v) fetal bovine serum (FBS), 1 mM L-glutamine, 100 U/mL penicillin, and 100 μM streptomycin.
2. Coimmunoprecipitation (co-IP) buffer: 50 mM HEPES, pH 7.4, 120 mM NaCl, 5 mM EDTA, 10% (v/v) glycerol, 1% (v/v) Triton X-100, 5 mM NaF, 1 mM sodium orthovanadate, 10 μg/mL benzamidine, 0.1 mM PMSF, 10 μg/mL soybean

trypsin inhibitor, 2% (w/v) EDTA-free complete protease inhibitor cocktail. The buffer minus protease inhibitors can be stored at 4 °C for several months. On the day of the experiment, the volume of buffer required can be taken from the stock and protease inhibitors added fresh from concentrated stocks (*see* **Note 1**).

3. Anti-Flag M2 affinity agarose gel suspended in buffered saline containing azide as preservative and 50% (v/v) glycerol.
4. 1-mL syringe to which a 25-gauge needle is attached.

2.2 Peptide Pull-Down

1. Human embryonic kidney (HEK) 293 cells maintained in Dulbecco's modified Eagle's medium (DMEM) supplemented with 10% (v/v) fetal bovine serum (FBS), 1 mM L-glutamine, 100 U/mL penicillin, and 100 μM streptomycin.
2. Coimmunoprecipitation buffer (as earlier).
3. N-terminally biotinylated peptides (Severn Biotech) dissolved in sterile purified water of pH 5–7 (*see* **Note 2**). Peptides are stable for several years when aliquoted and stored at −80 °C.
4. Streptavidin-Sepharose High-Performance beads in suspension with 20% (v/v) ethanol. Can be stored at 4 °C for several months; however, ensure that the buffer does not evaporate over time (*see* **Note 3**).
5. 1-mL syringe to which a 25-gauge needle is attached.

2.3 Peptide Array Overlay

1. Peptide array buffer: 50 mM HEPES, pH 7.4, 150 mM NaCl, 5 mM EDTA, 1% (v/v) Triton X-100, 5 mM NaF, 1 mM sodium orthovanadate, 10 μg/mL benzamidine, 0.1 mM PMSF, 10 μg/mL soybean trypsin inhibitor, 2% (w/v) EDTA-free protease inhibitor cocktail.
2. TBS-T: 50 mM Tris of pH 7.5, 150 mM NaCl, 0.1% (v/v) Tween 20.
3. Blocking buffer: TBS-T, 5% (w/v) BSA.
4. Binding buffer: TBS-T, 0.5% (w/v) BSA.
5. Purified recombinant Trx-polyHis-tagged SOCS3 (Sino Biological Inc., 11315-H30E) diluted in TBST-5% (w/v) BSA.
6. Primary antibody: goat anti-SOCS3 (M20: Santa Cruz, sc-7009). The antibody is stable for several years at 4 °C.
7. Secondary antibody: IRDye® 680LT Donkey anti-goat IgG (Licor, 925-68024).
8. 4-well dishes. Only required if using peptide arrays fabricated on a glass slide support.
9. Layflat tubing.
10. Heat sealer.

3 Methods

3.1 Coimmuno-precipitation

Coimmunoprecipitation (co-IP) is a simple technique that enables protein–protein interactions to be investigated. Target protein-specific antibodies conjugated to a bead support are incubated with soluble cell lysates allowing the formation of an immune complex between the bound target protein and its binding partner. The precipitated immune complex is then washed several times to remove nonspecifically bound proteins. Protein complexes are finally eluted from the beads prior to analysis by SDS-PAGE and immunoblotting.

The buffers used during coimmunoprecipitation must preserve protein–protein interactions and so a nonionic detergent such as Triton X-100 should ideally be used since it is less likely to disrupt preassembled protein complexes. Isolation of the immune complex also requires gentle handling of the sample from cell lysis through to elution. Sonication or vigorous vortexing during lysis and harsh resuspension of the precipitated immune complex during wash stages should be avoided. Instead, gentle pipetting and end-over-end mixing should be used so as not to disrupt protein-protein interactions.

This protocol describes the use of anti-Flag M2 antibody-conjugated affinity gel to coimmunoprecipitate Flag epitope (DYKDDDDK)-tagged SOCS3 and myc epitope (EQKLISEEDL)-tagged cavin-1 proteins from cotransfected HEK293 cells. To control for nonspecific binding to beads, we include reactions from cells expressing myc epitope-tagged cavin-1 alone. Cells expressing Flag epitope-tagged SOCS3 alone and cells expressing neither protein are used as positive and negative immunoprecipitation controls, respectively. Our data demonstrate that SOCS3 and cavin-1 coprecipitate and that the interaction is dependent on the presence of a PEST motif located in the SOCS3 SH2 domain (Fig. 1b).

Unless otherwise stated, all steps should be performed on ice.

1. Place 6-cm dishes of confluent HEK293 cells transiently overexpressing proteins of interest onto the ice and wash cell monolayers twice with 2 mL of ice-cold PBS using a sterile serological pipette. To avoid disrupting the cells, tilt dish and add the PBS the side of the dish, gently rock the dish to cover the cells, transfer spent PBS to waste with a pipette, and repeat. Remove residual PBS using a P1000.
2. Transfer 0.4-mL ice-cold PBS to each dish and scrape cells using a sterile cell scraper. Transfer cells to prechilled 1.5-mL microcentrifuge tubes using a P1000 (*see* **Note 4**).
3. Pellet the cells for 5 min at 600 × g using a refrigerated microcentrifuge set at 4 °C.

4. Remove PBS supernatant with a P200 without disturbing the cell pellet. At this stage, pellets can be stored at −20 °C.
5. Gently resuspend pellet in 50–150 μL freshly made co-IP buffer supplemented with protease inhibitors using a P200 (*see* **Note 5**).
6. Mix samples end-over-end *via* rotation for 30 min at 4 °C using a rotating mixer placed in a 4 °C cold room or chiller cabinet.
7. Pellet detergent-insoluble cellular material by centrifugation for 15 min at 21,000 × *g* using a refrigerated microcentrifuge set at 4 °C.
8. Isolate supernatant to fresh prechilled 1.5-mL microcentrifuge tubes.
9. Determine the protein concentration in each sample using a bicinchoninic acid (BCA) protein assay. We typically use 0.5–1 mg per co-IP reaction and keep back 10–20 μg of each sample as protein expression controls for running in parallel to the co-IP samples.
10. Prepare the precipitating anti-Flag M2 antibody-conjugated affinity gel. The gel provided as a 50% (v/v) slurry suspended in 50% (v/v) glycerol and needs to be washed thoroughly before use. Aliquot the required total volume of 50% slurry in a single microcentrifuge tube and wash three times with 1 mL of co-IP buffer (*see* **Note 6**). We use 30 μL of 50% (v/v) slurry of gel per reaction. Completely resuspend the gel in co-IP buffer by end-over-end mixing and pellet by centrifugation for 2 min at 300 × *g* using a refrigerated microcentrifuge set at 4 °C between washes. Remove co-IP buffer using a P1000 pipette without disturbing the pellet. After the final wash stage, completely remove any excess co-IP buffer with a 1-mL syringe to which a 25-gauge needle is attached.
11. Resuspend the gel in co-IP buffer to make a 50% (v/v) slurry.
12. A liquot gel as required into empty prechilled microcentrifuge tubes using a P200 with 3 mm cut off the tip to prevent blockage by the gel matrix (*see* **Notes 6–10**).
13. Transfer 0.5–1 mg of soluble protein lysates directly to the aliquoted gel, close, and secure lids with Nescofilm.
14. Mix end-over-end using a rotating mixer overnight at 4 °C.
15. Isolate the immune complex by pelleting by centrifugation for 2 min at 300 × *g* using a refrigerated microcentrifuge set at 4 °C.
16. Remove the unbound soluble protein lysate to fresh prechilled microcentrifuge tubes using a P1000. Do not disturb the gel, leave a small volume of soluble protein lysate 10–20 μL, which can be removed during the wash stages. These samples can be

analyzed *via* SDS-PAGE in parallel with co-IPs to assess the percentage of precipitated protein capture for troubleshooting purposes.

17. Wash the gel-bound immune complex three times each with 1 mL co-IP buffer to remove nonspecifically bound proteins. At this stage, it is essential that the samples are handled with care. Gently drip the co-IP buffer down the side of the microcentrifuge tube and carefully mix end-over-end to resuspend. Pellet by centrifugation for 2 min at 300 × *g* using a refrigerated microcentrifuge set at 4 °C before repeating the wash.
18. After the final wash, remove residual co-IP buffer with a 1-mL syringe to which a 25-gauge needle is attached.
19. Elute immune complexes from the gel by addition of 45 μL 4× Laemmli sample buffer supplemented with dithiothreitol (DTT) to a final concentration of 1 mM and heat in a heat block or water bath set at 95 °C for 5 min (*see* **Note 11**).
20. Following centrifugation for 1 min at 1000 × *g* at room temperature, transfer the eluate to fresh room temperature 1.5-mL microcentrifuge tubes using a fine-gauge Hamilton syringe, washing the syringe ten times with distilled deionized water between each sample. At this stage, samples can be stored at −20 °C.
21. Analyze 20 μL of eluate *via* SDS-PAGE and store the remaining volume at −20 °C as a backup (*see* **Note 11**).

3.2 Peptide Pull-Down

Peptide pull-downs are similar to coimmunoprecipitation experiments in that protein complexes are captured. However, instead of a target-specific antibody, a short peptide from a protein of interest is used as bait to bind desired target proteins. Peptides can be modified (biotinylated) allowing noncovalent high-affinity (K_d for streptavidin-biotin interaction is in the order of 10^{-14} M) conjugation to a support gel (streptavidin-Sepharose), which is then incubated with soluble protein lysates prior to precipitation and analysis by SDS-PAGE and immunoblotting [8].

This protocol describes the use of streptavidin-Sepharose beads and an N-terminal biotin-conjugated peptide containing the SOCS3 interaction motif from the gp130 receptor (PTyr759 (pY), biotin-TSSTVQpYSTVVHSG) to precipitate SOCS3. To control for nonspecific binding, we include reactions using unmodified (nonphosphorylated) and scrambled peptide sequences. Our data demonstrate that SOCS3 is specifically precipitated by the tyrosine-phosphorylated gp130 peptide and that the interaction is not dependent on the presence of the SH2 domain-localized PEST motif (Fig. 1c).

Unless otherwise stated, all steps should be performed on ice.

1. Place 6-cm dishes of confluent HEK293 cells transiently over-expressing proteins of interest onto ice and carefully wash cell monolayers twice with 2 mL of ice-cold PBS using a sterile serological pipette. To avoid disrupting the cells, tilt the dish and add the PBS to the side of the dish, gently rock to cover the cells, transfer spent PBS to waste with a pipette, and repeat. Remove residual PBS using a P1000.
2. Transfer 0.4-mL ice-cold PBS to each dish and scrape cells using a sterile cell scraper. Transfer cells to prechilled 1.5-mL microcentrifuge tubes using a P1000 (*see* **Note 4**).
3. Pellet the cells for 5 min at 600 × *g* using a refrigerated microcentrifuge set at 4 °C.
4. Remove PBS supernatant with a P200 without disturbing the cell pellet. At this stage, pellets can be stored at −20 °C.
5. Gently resuspend pellet in 50–150 μL freshly made co-IP buffer supplemented with protease inhibitors using a P200 (*see* **Note 5**).
6. Mix samples end-over-end *via* rotation for 30 min at 4 °C using a rotating mixer placed in a 4 °C cold room or chiller cabinet.
7. Pellet detergent-insoluble material by centrifugation for 15 min at 21,000 × *g* using a refrigerated microcentrifuge set at 4 °C.
8. Isolate supernatant to fresh prechilled 1.5-mL microcentrifuge tubes; be sure not to transfer over any material from the pellet.
9. Determine the protein concentration in each sample. We use a bicinchoninic acid (BCA) protein assay and typically require 0.5–1 mg per co-IP reaction and keep back 10–20 μg per sample as protein expression controls.
10. Prepare the precipitating streptavidin-Sepharose beads. The beads are provided as a 50% (v/v) slurry in 20% (v/v) ethanol and need to be washed thoroughly before use. Aliquot the required total volume of 50% (v/v) slurry in a single micro-centrifuge tube and wash three times with 1 mL of co-IP buffer (*see* **Note 6**). We use 30 μL of 50% (v/v) slurry of beads per reaction. Completely resuspend the beads in co-IP buffer by end-over-end mixing and pellet by centrifugation for 2 min at 300 × *g* using a refrigerated microcentrifuge set at 4 °C between washes. Remove co-IP buffer using a P1000 pipette without disturbing the pellet. After the final wash stage, completely remove any excess co-IP buffer with a 1-mL syringe to which a 25-gauge needle is attached.
11. Resuspend the beads in co-IP buffer to make a 50% (v/v) slurry.

12. Aliquot beads as required into empty prechilled microcentrifuge tubes using a P200 with 3-mm cut-off the tip to prevent blockage by beads (*see* **Notes 6–10**).
13. To reduce background from proteins that bind nonspecifically to the beads, preclear the soluble extracts by incubating with the unconjugated streptavidin-Sepharose beads. Prepare 0.5–1 mg of soluble protein lysates in a total 1 mL of co-IP buffer and add to an aliquot of beads.
14. Mix end-over-end using a rotating mixer for 1 h at 4 °C.
15. Isolate the precleared soluble protein lysates following centrifugation for 5 min at 300 × *g* using a refrigerated microcentrifuge set at 4 °C.
16. Transfer soluble extracts directly to a fresh aliquot of streptavidin-Sepharose beads in prechilled 1.5-mL microcentrifuge tubes.
17. Add an optimized concentration of precipitating peptide (100 nM) to each pull-down reaction, close, and secure lids with Nescofilm. To control for nonspecific binding, perform parallel pull-downs using 100 nM of unmodified and scrambled control peptides.
18. Mix end-over-end using a rotating mixer for 16 h at 4 °C.
19. Isolate peptide-protein complexes by pelleting by centrifugation for 2 min at 300 × *g* using a refrigerated microcentrifuge set at 4 °C.
20. Remove the unbound soluble protein lysate to fresh prechilled microcentrifuge tubes using a P1000. Do not disturb the beads, leave a small volume of soluble protein lysate 10–20 μL, which can be removed during the wash stages. These samples can be analyzed *via* SDS-PAGE in parallel with pull-downs to assess the percentage of precipitated protein capture for troubleshooting purposes.
21. Wash the bead-bound peptide-protein complex three times with 1-mL co-IP buffer to remove nonspecifically bound proteins. At this stage, it is essential that the samples are handled with care. Gently drip the co-IP buffer down the side of the microcentrifuge tube and carefully mix end-over-end to resuspend. Pellet by centrifugation for 2 min at 300 × *g* using a refrigerated microcentrifuge set at 4 °C before repeating.
22. After the final wash, remove excess co-IP buffer with a 1-mL syringe to which a 25-gauge needle is attached.
23. To elute the protein complex from the beads, add 45 μL 4× Laemmli sample buffer supplemented with 1 mM DTT and heat in a heat block or water bath set at 95 °C for 5 min.

22. Transfer the eluate to fresh room temperature 1.5-mL microcentrifuge tubes using a fine-gauge Hamilton syringe, washing the syringe ten times with distilled deionized water between each sample. At this stage, samples can be stored at −20 °C.

24. Analyze 20 μL of eluate *via* SDS-PAGE and store the remaining volume at −20 °C as back-up.

3.3 Peptide Array Overlay

Peptide arrays offer a high-throughput method of investigating direct protein-protein interactions (Fig. 2). Through the binding of hundreds to thousands of peptides to a cellulose support, analytical SPOT synthesis allows for detailed investigation of interactions with other peptides/proteins of interest. For our study, peptides of interest were synthesized using 9-fluorenylmethyloxycarbonyl chemistry (f-moc) by the Autospot-robot ASS 222 (Intavis Bioanalytical Instruments). Peptide microarrays consisting of overlapping 25mer sequences spanning the full length of the protein of

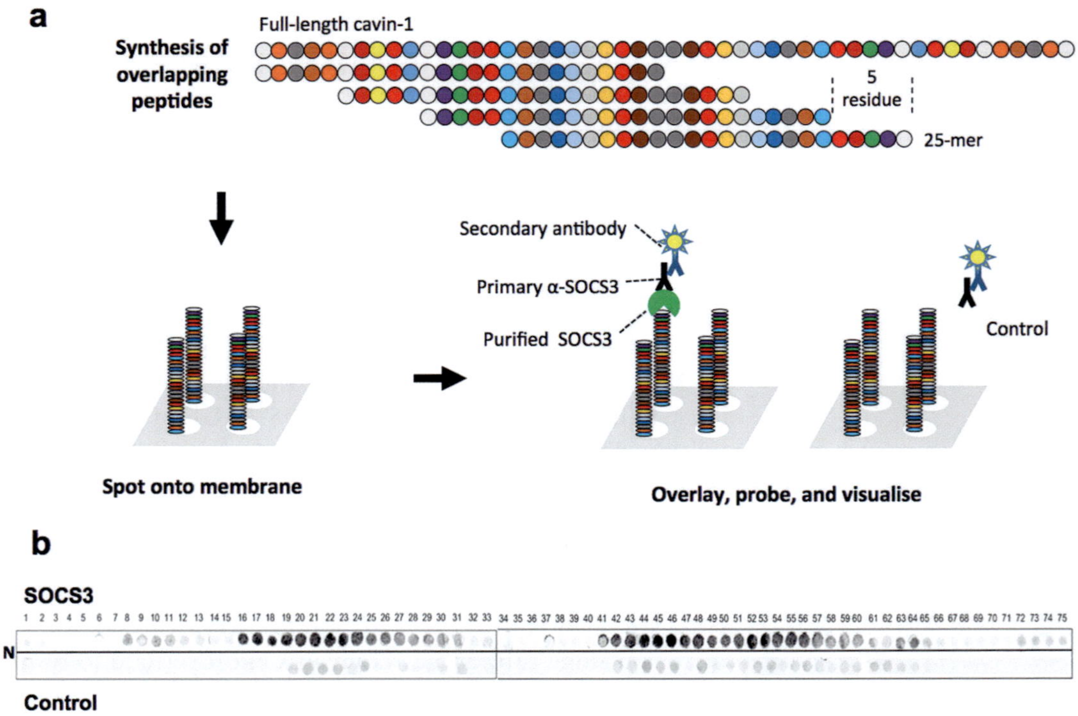

Fig. 2 Interrogation of SOCS3 interaction sites in cavin-1 using an overlapping cavin-1 peptide array. (**a**) Using an automated system, 25mer, 5-residue overlapping cellulose-conjugated peptides spanning the cavin-1 primary sequence from the N- to C-terminus are synthesized and spotted onto a support layer producing a one- or three-dimensional layer of peptides. Following overlay with purified recombinant SOCS3 (SOCS3) or buffer without SOCS3 (Control), interactions are detected by sequential incubation with a SOCS3-specific antibody and a secondary antibody. (**b**) Representative data from overlays with (SOCS3) and without (Control) purified SOCS3. Dark spots represent areas of interaction between SOCS3 and peptides within the cavin-1 array. The locations of N- and C-terminal peptides are indicated

interest, sequentially shifted by 5 amino acids, were covalently bound to the cellulose support by their C-terminal residue [9, 10].

Since peptides are synthesized de novo, common post-translational modifications (e.g., phosphorylation), point mutations to assess interaction interfaces (e.g., alanine-scanning approaches), and disease-relevant variations can be easily generated enabling important regions/residues to be identified. The peptide array is overlaid with target protein either purified or expressed in soluble protein lysates from transfected cells, and interactions detected by probing with a target protein-specific antibody. Peptide arrays can be produced on loose cellulose membranes or membranes supported on glass slides [11]. This is a general protocol that can be used with either support, with any variations indicated in the notes.

This protocol describes the interrogation of the SOCS3/cavin-1 interaction using a full-length human cavin-1 peptide array, synthesized as 25mer, 5-residue overlapping cellulose-conjugated peptides overlaid with purified recombinant SOCS3 in its native state. Interactions are detected using a SOCS3-specific antibody followed by a conjugated secondary antibody. To control for background, peptide arrays are treated with primary and secondary antibodies only. Our data demonstrated that SOCS3 specifically interacts with two regions within cavin-1 (Fig. 2).

1. Equilibrate the peptide arrays in TBS-T for 10 min at room temperature (*see* **Notes 12–14**).
2. Block in 3 mL of TBS-T containing 5% (w/v) BSA for 4 h at room temperature to prevent nonspecific protein binding (*see* **Note 15**).
3. Peptide arrays were overlaid with either purified SOCS3 protein (10 μg/mL) in a total volume of 2.5 mL with TBST containing 0.5% (w/v) BSA and protease inhibitors (10 μg/mL benzamidine, 0.1 mM PMSF, 10 μg/mL soybean trypsin inhibitor, 2% (w/v) EDTA-free complete protease inhibitor cocktail) overnight at 4 °C with rocking (*see* **Notes 16–18**).
4. Wash the array three times for 10 min in 2.5 mL TBS-T with rocking.
5. Incubate the peptide array with the primary anti-SOCS3 (goat) antibody diluted in a total volume of 2.5 mL TBS-T containing 1% (w/v) BSA overnight at 4 °C with rocking (*see* **Notes 19–21**).
6. Wash the array three times for 10 min in 2.5 mL TBS-T with rocking.
7. Incubate the peptide array with the IRDye® 680LT Donkey anti-goat IgG secondary antibody diluted in a total volume of 2.5 mL TBS-T containing 1% (w/v) BSA for 45 min at room temperature (*see* **Note 22**)

8. Wash the array three times for 10 min in 2.5 mL TBS-T with rocking.
9. Visualize using preferred gel documentation system.

4 Notes

1. Stocks of soybean trypsin inhibitor and benzamidine (each 10 mg/mL) can be made up in deionized water and stored in single-use aliquots at −20 °C for at least 6 months. PMSF must be made up fresh from powder to a concentration of 10 mM in methanol on the day of the experiment before being added to the appropriate buffer to produce a final concentration of 0.1 mM.
2. The biochemical properties of the peptide dictate the solvent needed to dissolve the peptide. As such, supplier information should be referenced to optimize this stage.
3. To minimize evaporation once opened, stock bottles of protein G–Sepharose beads should be capped tightly and the seal sheathed with tightly wrapped Nescofilm after each use. However, should significant evaporation occur over time, make up 20% (v/v) ethanol in PBS and add enough of this solution to the bottle so that the ratio of bead to buffer volume is maintained at approximately 1:1.
4. While cells can be scraped directly into co-IP buffer, this step enables the volume of lysis buffer required to be estimated from the pellet size and avoids over dilution of the protein sample.
5. To preserve protein–protein interactions, do not vortex or sonicate during or after this stage.
6. The pipette tip can sometimes become blocked by the beads, cut a few millimeters off the top of the tip to enlarge the hole.
7. Aliquoting into empty tubes will allow a comparison of aliquots and avoids large variations in bead volume.
8. If a large number of reactions are required, frequently resuspend bead stock before aliquoting.
9. If more than 400 μL of 50% slurry is required, distribute between multiple microfuge tubes.
10. To account for pipetting errors, prepare one reaction volume extra than required.
11. Protein complexes and the precipitating antibody will be present in the eluate when using Laemilli sample buffer and DTT. Antibodies will be denatured and reduced to heavy (~50 kDa) and light (~25 kDa) chains, which might mask proteins of interest following immunoblotting. To avoid this, we use

protein G-conjugated HRP (Millipore), which only binds to nondenatured antibodies, instead of a secondary HRP-conjugated antibody.

12. Loose cellulose membrane should be rinsed in 100% ethanol prior to TBS-T wash.
13. We have used peptide arrays fabricated on Whatman® 50 cellulose membrane or membranes supported on glass slides as in CelluSpot® [11] peptide arrays. The latter has a three-dimensional layer of peptides, which increases binding capacity and improves detection of weak interactions by providing a greater number of peptides in a given area. However, peptide arrays supported on glass slides are sensitive to mechanical stress and so to address this limitation and to reduce volumes of solutions used, all incubations were performed in 4-well dishes.
14. Loose cellulose membranes can be washed in plastic tubs or tubes as required. To reduce the volume of all solutions, plastic pouches made from heat-sealed layflat tubing can be used for overlaying the peptide array with protein or incubation with antibodies. It is important to remove any bubbles before sealing the pouch so that the array is always covered by the solution. Volumes required will depend on the size of the array; typical volumes range from 2 to 4 mL.
15. If your array contains phosphopeptides, it is recommended to use PhosphoBLOCKER (Cell Biolabs Inc.) in place of milk or BSA.
16. If you are overlaying the peptide array with cell lysates, if required, add phosphatase inhibitors to preserved protein phosphorylation.
17. Prepare cell lysates to be highly concentrated by lysing cells in a small volume, i.e., 250 μL of peptide array buffer per 10-cm-diameter dish.
18. You will need to optimize the amount of protein needed to detect specific interactions. Overlay the peptide array with several dilutions of your purified protein, e.g., 10, 1, 0. 1, 0.01 μg, if using cell lysates more will be required, e.g., 1000, 500, 250, 100 μg. Probe the array with a specific primary antibody and compare with a nonoverlaid control or untransfected cell lysate. If your protein is highly overexpressed, then not much lysate will be required. If the slide comes up blank, incubate with a higher protein concentration without stripping the array.
19. Primary antibody concentration will need to be optimized for your application. Typical concentrations range from 1:10 to 1:1000.

20. Alternatively, the peptide array can be incubated with the primary antibody for 3–8 h at room temperature.
21. Many different secondary-tagged reporter groups are available, perhaps the most common being HRP and fluorescent dyes. Due to their wide dynamic range, we have found fluorescent dyes to improve the detection of weak interactions. Due to the sensitivity of these antibodies to light, incubation periods are performed in the dark.
22. Secondary antibody concentration will need to be optimized for your application. Typical concentrations range from 1:1000 to 1:10,000.

References

1. Linossi EM, Babon JJ, Hilton DJ et al (2013) Suppression of cytokine signaling: the SOCS perspective. Cytokine Growth Factor Rev 24:241–248
2. Williams J, Munro K, Palmer T (2014) Role of ubiquitylation in controlling suppressor of cytokine signalling 3 (SOCS3) function and expression. Cell 3:546–562
3. Babon JJ, Varghese LN, Nicola NA (2014) Inhibition of IL-6 family cytokines by SOCS3. Semin Immunol 26:13–19
4. Kershaw NJ, Laktyushin A, Nicola NA et al (2014) Reconstruction of an active SOCS3-based E3 ubiquitin ligase complex in vitro: identification of the active components and JAK2 and gp130 as substrates. Growth Factors 32:1–10
5. Zhang WN, Wang L, Wang Q et al (2012) CUEDC2 (CUE domain-containing 2) and SOCS3 (suppressors of cytokine signaling 3) cooperate to negatively regulate Janus kinase 1/signal transducers and activators of transcription 3 signaling. J Biol Chem 287:382–392
6. Williams JJ, Alotaiq N, Liu L et al (2018) Suppressor of cytokine signaling 3 (SOCS3) interaction with cavin-1 links SOCS3 function and cavin-1 stability. Nat Commun 9:168
7. Liu L, Brown D, McKee M et al (2008) Deletion of Cavin/PTRF causes global loss of caveolae, dyslipidemia, and glucose intolerance. Cell Metab 8:310–317
8. Chaiet L, Wolf FJ (1964) The properties of streptavidin, a biotin-binding protein produced by Streptomycetes. Arch Biochem Biophys 106:1–5
9. Bolger GB, Baillie GS, Li X et al (2006) Scanning peptide array analyses identify overlapping binding sites for the signalling scaffold proteins, β-arrestin and RACK1, in cAMP-specific phosphodiesterase PDE4D5. Biochem J 398:23–36
10. Reijns MAM, Bubeck D, Gibson LCD et al (2011) The structure of the human RNase H2 complex defines key interaction interfaces relevant to enzyme function and human disease. J Biol Chem 286:10530–10539
11. Wu C, Li SS-C (2009) CelluSpots™: a reproducible means of making peptide arrays for the determination of SH2 domain binding specificity. Meths Mol Biol 570:197–202

Chapter 11

Analysis of Protein and Lipid Interactions Using Liposome Co-sedimentation Assays

Elin Larsson, Madlen Hubert, and Richard Lundmark

Abstract

The dynamic assembly of proteins at the membrane interphase is key to many cell biological processes such as the generation and stabilization of caveolae at the cell surface via coat proteins. The liposome co-sedimentation assay has been widely used for studies of protein and lipid interactions and has provided important information about binding mechanisms, lipid-binding specificity, and curvature preference of proteins. Here, we describe this technique in detail and how it can be used as a tool to address the membrane-binding ability and lipid specificity of caveolae-associated proteins.

Key words Co-sedimentation, Liposome pull down, Protein and lipid interactions, Liposomes, SUVs, Lipid specificity, EHD2, Cavin, Caveolae coat

1 Introduction

A detailed analysis of protein–lipid interactions is key to our understanding of transient protein assembly at the membrane interphase such as the formation of the caveolae coat. Co-sedimentation analysis of purified proteins together with small unilamellar vesicles (SUVs) has proven to be a potent tool for such studies [1–6]. This technique has been widely used for studying protein and lipid interactions and, in combination with protein mutagenesis, it has provided key information about lipid binding specificity of proteins as well as membrane curvature preference [1, 2, 4, 7–9]. The assay allows studying the direct interactions between defined lipid species and purified protein components. The ability to precisely define the membrane composition of the SUVs enables systematic analysis of the lipid species involved in protein assembly. Proteins are incubated together with SUVs to facilitate binding after which the SUVs are pelleted by centrifugation and proteins that are bound will co-sediment with the SUVs, while unbound protein will remain in the supernatant. Analysis of the protein content in the supernatant and the pellet fractions using

Cedric M. Blouin (ed.), *Caveolae: Methods and Protocols*, Methods in Molecular Biology, vol. 2169,
https://doi.org/10.1007/978-1-0716-0732-9_11,

SDS-PAGE reveals the percentage of protein that interacts with membranes at these conditions. Here, we describe how this technique has been used to address the membrane interaction of the caveolae proteins EHD2 and cavins. We have employed this technique to study the lipid binding of cavin1 and cavin3 and its dependence on cholesterol levels (Figs. 1 and 2) [4]. Using this technique, we have also been able to characterize the amino acids in EHD2 that interacts with lipids, the specific dependence of nucleotides for binding, and the regulatory role of specific domains (Fig. 3) [2, 10].

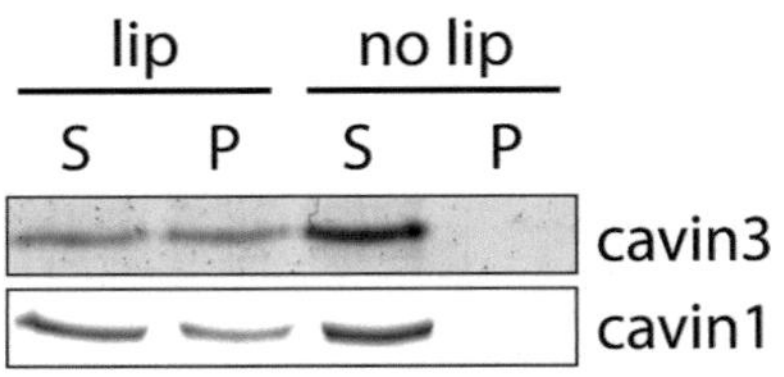

Fig. 1 Cavin3 and cavin1 associates and co-sediments with liposomes. Full-length cavin3 or cavin1 were incubated with Folch liposomes (lip) or without (no lip) before centrifugation. Supernatant (S) and pellet (P) fractions were separated by SDS-PAGE and visualized using Coomassie brilliant blue stain. (Reproduced from ref. 4 with permission from The company of Biologists Ltd)

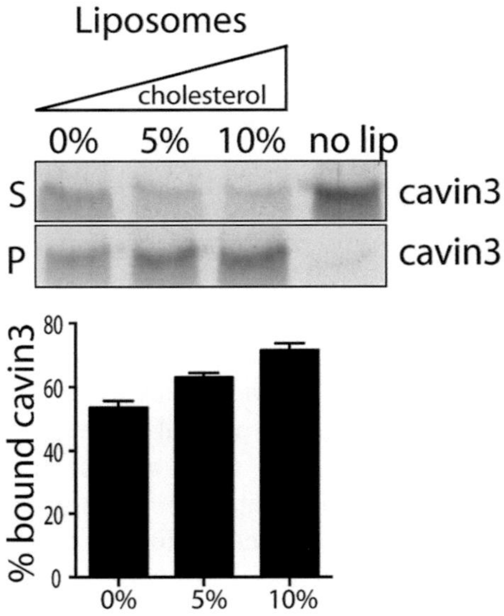

Fig. 2 Cavin3 association to Folch SUVs is increased with elevated cholesterol concentration. Cholesterol was mixed with total brain lipids (Folch fraction) to obtain liposomes that contained either 5 or 10% cholesterol. Supernatant (S) and pellet (P) fractions were separated by SDS-PAGE and visualized by Coomassie brilliant blue stain. Bar graph shows the quantification of pelleted cavin3. The values represent mean ± s.d. from three independent experiments. (Reproduced from ref. 4 with permission from The company of Biologists Ltd)

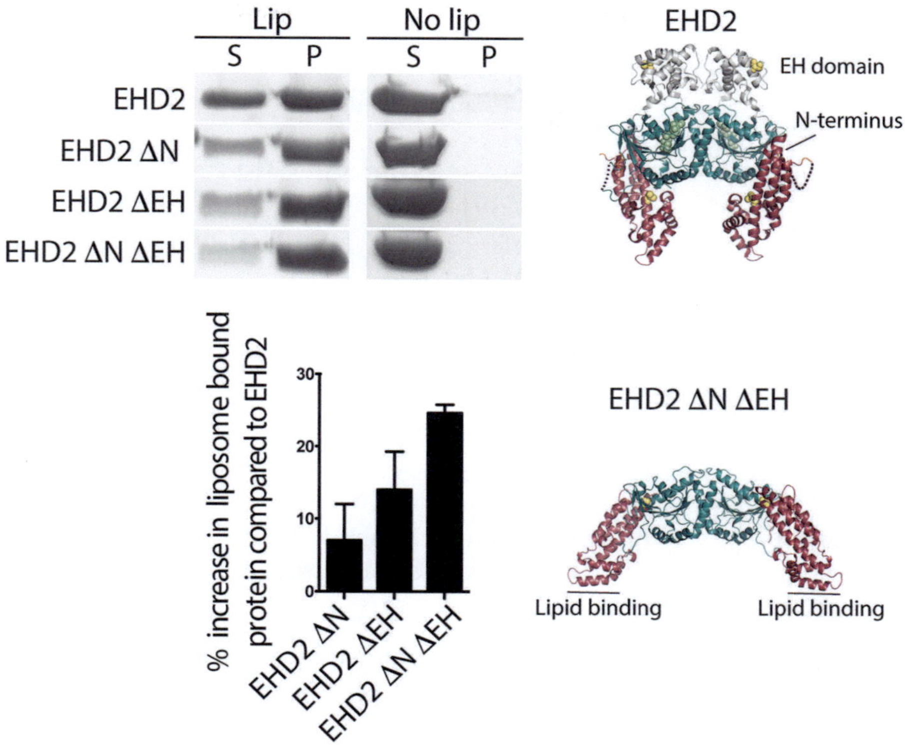

Fig. 3 The N terminus and EH domain influence the membrane association of EHD2. (**a**) Different EHD2 protein variants (2.5 μM) loaded with AMP-PNP (1 mM) was incubated with Folch liposomes (0.5 mg/mL) before centrifugation. Supernatant (S) and pellet (P) fractions were separated by SDS-PAGE and visualized by Coomassie brilliant blue stain. The bar graph shows the quantification of pelleted EHD2 variants. The values represent mean ± s.d. from three independent experiments. (**b**) Structure representation of full-length EHD2 in autoinhibited state and N terminus and EH-domain EHD2 deletion mutant primed for membrane binding. (Reproduced from ref. 10 with permission from National Academy of Sciences)

2 Materials

2.1 Lipids

1. Lipids are dissolved in either chloroform or a mixture of chloroform/methanol (3:1, v/v) (*see* **Note 1**). Store the lipid stock solutions in glass vials with screw caps including a Teflon-covered liner (*see* **Note 2**) layered with nitrogen at −20 °C.
2. Total brain lipids (Folch fraction) and other common lipids (*e.g.*, glycerolphospholipid, sphingolipid, sterol) can be used to generate customized types of liposomes (*see* **Note 3**).

2.2 Liposome Preparation

1. Lipid stock solutions.
2. Rehydration buffer: 20 mM HEPES (pH 7.4), 150 mM NaCl.
3. Glass vials (1.5 mL).
4. Hamilton microliter syringes.
5. Dry nitrogen stream.

6. Vortex mixer.
7. Bath sonicator.
8. Glass beads (diameter 1 mm).

2.3 Protein

1. EHD2, cavin1 or cavin3 expressed and purified from Rosetta DE cells [4, 10].
2. Protein buffer: 20 mM HEPES-NaOH (pH 7.4), 150 mM NaCl, 1 mM $MgCl_2$.
3. Nucleotide derivates: ADP, ATP, ATP-δS or AMP-PNP. Stock solution 100 mM: dissolved in 200 mM $MgAc_2$ and neutralized with NaOH.
4. Refrigerated centrifuge.
5. Eppendorf tubes (1.5 mL).

2.4 Centrifugation

1. Beckman Coulter Optima™ MAX (or MAX-E, TL, TLX and TL-100) tabletop. Ultracentrifuge.
2. Ultracentrifuge rotor: Type TLA 100.
3. Polycarbonate centrifuge tubes (7 × 20 mm, 0.2 mL).
4. Suitable tube rack (7 mm, tubes).
5. Sample buffer (6×SB): 10 mL 1 M Tris–HCl (pH 7.4), 6 g SDS, 30 mL glycerol, 10 mL β-mercaptoethanol. Add a few grains of bromophenol blue. Keep working aliquot at 4 °C and store the remaining aliquots at −20 °C. Use 3×SB for assay (*see* **Note 4**).
6. Eppendorf tubes (1.5 mL).
7. Forceps.

2.5 SDS-PAGE Analysis

1. Precast SDS-Polyacrylamide gels 12.5%.
2. Electrophoresis cell.
3. Electrophoresis buffer: 49.5 mM Tris Base, 383 mM Glycine, 1% SDS.
4. Power Pac.
5. Microloading tips.
6. Prestained protein ladder.
7. Coomassie brilliant blue solution: Coomassie brilliant blue dissolved in destain solution.
8. Destain solution: 55% ethanol, 10% acetic acid, 35% distilled water.
9. ChemiDoc Imaging system.
10. Image Lab Software.

3 Methods

3.1 Liposome Preparation

1. Prepare liposomes at a total lipid concentration of 1 mg/mL. Mix chloroform and methanol (3:1, v/v) in a small glass vial (1.5 mL) and add the lipids in the appropriate ratio. Use Hamilton microliter syringes rather than plastic pipettes for pipetting organic solvents (*see* **Note 2**). Vortex gently to homogenize the lipid mixture.
2. Carefully evaporate the organic solvent by slowly rotating the vial under light nitrogen stream to yield a lipid film (*see* **Note 1**). To remove residual organic solvents and thoroughly dry the lipid film, keep the vial under a stronger nitrogen stream for another 30 min.
3. Rehydrate the lipid film with an appropriate volume of rehydration buffer to obtain a final concentration of 1 mg/mL. Glass beads can be added to help detach the lipid film from the inner glass surface. The mixture is then vigorously agitated using a vortex mixer and is allowed to hydrate at room temperature for 30 min (*see* **Note 5**).
4. To reduce the size of the liposome, place the vials containing the lipid dispersion in a bath sonicator (*see* **Note 5**) and treat the sample with five cycles of sonication (1 min on/1 min off).
5. Keep liposomes at room temperature if proceeding with the assay or store at 4 °C for a maximum of 2 days.

3.2 Protein Preparation

1. Gently thaw the protein on ice and centrifuge at 20,000 × *g* for 10 min at 4 °C to remove any precipitate. Carefully transfer soluble protein to a new, prechilled Eppendorf tube and keep it on ice (*see* **Note 6**).
2. Determine the protein concentration and dilute to 4 μM in protein buffer and keep it on ice (*see* **Note 3**).
3. EHD2 is preincubated with 1 mM nucleotide (ADP, ATP, ATP-δS, or AMP-PNP) for 10 min at room temperature to prime it for lipid binding. Proceed immediately with co-sedimentation assay. The cavin proteins do not require pretreatment with nucleotide.

3.3 Co-sedimentation

1. Label and place polycarbonate centrifuge tubes in a suitable tube holder. Include tubes for negative controls where rehydration buffer is used instead of liposomes.
2. The liposome concentration is within the range of 0.05–0.5 mg/mL (*see* **Notes 3** and **7**) in the final volume. Gently vortex the liposomes and prepare dilutions using rehydration buffer. Carefully add the desired volume to designated tubes by pipetting (*see* **Note 8**).

3. Add rehydration buffer to tubes designated for negative controls. Use the same volumes as for liposomes in **step 2**.
4. Resuspend the protein by pipetting and add to the centrifuge tubes in a 1:1 ratio to either liposomes or rehydration buffer. The final protein concentration in the tube is 2 μM (*see* **Note 8**).
5. Incubate protein/liposome or protein/buffer mixtures at room temperature for 10 min before centrifugation (*see* **Note 9**).
6. Centrifuge the samples at 100,000 × *g* for 20 min at room temperature (*see* **Notes 10** and **11**).
7. Remove the rotor carefully from the ultracentrifuge and take out the tubes from the rotor one at a time using forceps. Identify where the liposome pellet is located (*see* **Note 12**) and carefully collect the supernatant without disturbing the pellet (*see* **Note 13**). Resuspend the supernatant in a prelabeled tube containing an appropriate amount of 3×SB (*see* **Note 4**). Continue to collect the supernatant from all the samples and place the polycarbonate ultracentrifugation tubes in the tube holder and keep track of the placement of the pellet.
8. Resuspend the pellets in rehydration buffer using the same total volume as in **step 4** (*see* **Note 14**). Add the sample to a prelabeled tube containing an appropriate amount of SB (*see* **Note 4**).
9. Briefly centrifuge to collect the samples in the bottom of the tubes before and after boiling them at 95 °C for 10 min.

3.4 Analysis by SDS-PAGE

1. Load appropriate sample volumes into wells of a 12% SDS-polyacrylamide gel (*see* **Note 15**). The electrophoresis is run at a limit of 150 V for 60 min.
2. Stain the proteins by incubating the gel in Coomassie brilliant blue solution (*see* **Note 16**) on slow agitation until visible bands are showing in any of the lanes.
3. Remove excess color in the gel by incubating the gel in destain solution until only the proteins are blue.
4. Document the result by scanning the gel on an Imaging System and save it as a high-quality file (*see* **Note 17**).
5. Quantify the percentage of pelleted protein and protein remaining in the supernatant between the different samples (*see* **Note 18**) using an image processing software of choice.
6. If the open-source software ImageJ is used, files are opened and converted to gray scale by changing the image type to 32-bit.
7. The brightness and contrast of the image may be changed for optimal quantification.

8. Using the selection tool, create a rectangle around the protein band (*see* **Note 19**).
9. Go to "Analyze gel" then "Select first lane" and then physically drag the rectangle to the next protein band. Go to "Analyze gel" then "Select" and drag the box to the next protein band and repeat the latest step until all bands are selected.
10. Band intensities are plotted by selecting "Analyze gel" and "Plot lanes." To get the area under the peak, a straight line is drawn using the "Straight line tool" to mark off the peak. The process is repeated for all peaks (*see* **Note 20**).
11. Select each band peak by using the "Wand tool" and the value of the corresponding peak area will be shown in a new window. Add the value of the area of the pellet and the area of the supernatant to get the value of the total protein area (*see* **Note 21**). Calculate the ratios between pellet:supernatant for all samples by dividing the area of the pellet with the total area.
12. Repeat the quantification process at least three times and use the average value.

4 Notes

1. Wear protective gloves when handling organic solvents and working in a fume hood.
2. We recommend performing the experiments with glassware only to avoid any source of contamination by greases and plasticizers from plastics and detergents.
3. To study protein–lipid interactions, it is crucial to consider parameters such as lipid composition, molar ratio of the lipids, membrane curvature, pH, and salt concentration of the buffer.
4. We usually add 25 μL of a 3×SB to a final sample volume of 50 μL. We never use a more concentrated SB than 3× since a more concentrated buffer can be difficult to pipet and can distort the final volumes between the different samples.
5. It is important to consider the phase transition of the employed lipids. To ensure that the lipids are in the fluid phase during liposome formation, the rehydration buffer should be above the phase transition temperature (T_m) of the lipid with the highest T_m.
6. Avoid freeze-thawing the protein. If the protein stock is to be used again shortly, it is better to keep it on ice at 4 °C, rather than to snap freeze it in liquid nitrogen and storing it at −80 °C.

7. We usually prepare a total volume of 50 μL per centrifuge tube as this volume does not require a lot of reagents but is still large enough to separate pelleted liposomes from non-pelleted protein.
8. To make sure that protein is mixed with the liposomes or buffer, we pipet the reagents close to the bottom of the tubes to avoid drops adhering to the walls or air pockets.
9. During the incubation, we carefully place the tubes in the ultracentrifugation rotor using forceps. Load the filled tubes symmetrically into the rotor. It is very important to properly attach the rotor in/to the ultracentrifuge to not cause any accidents or to damage the centrifuge. As it might take some time for the centrifuge to work up the vacuum, we recommend to start and set the parameters on the centrifuge so that the samples are ready for centrifugation once the incubation time is over.
10. We advise to make sure that the centrifuge reaches the desired speed before leaving it unattended. To avoid that the liposome pellet slips from the tube wall, we recommend to collect the samples as soon as the centrifugation is finished and proceed with the experiment directly.
11. While the samples are being centrifuged, we always prepare Eppendorf tubes to collect both supernatant and pellet from each sample. The tubes are labeled and 3×SB is added.
12. The pellets can be difficult to identify but are placed on the wall near the bottom of the tube, facing upward in the rotor. Shining light onto the tube can aid locating the pellet as it may be colorless. It can also be of help to mark the position of the pellet by adding a dot on the tube using a labeling pen.
13. To remove all supernatants, the pipet can be set to collect a greater volume than the volume in the tube.
14. Add the buffer just above the pellet and resuspend it by pipetting up and down going over the pellet. Make sure that the entire pellet is resuspended before adding it to the prepared Eppendorf tube.
15. We typically load 20 μL of each sample in a Mini-PROTEAN 15-well, 1.0 mm, which can hold a maximum of 26 μL.
16. The coloring efficiency of the staining solution can be controlled by adjusting the amount of Coomassie brilliant blue.
17. We typically save the file as TIFF, but the image could also be saved as a JPEG or GIF depending on which program is used for quantification.
18. We generally use ImageJ, which can be downloaded for free from: https://imagej.net/Downloads.

19. For easier quantification process, we encourage that the protein bands are as straight as possible and not tilted or skewed. Create a rectangle with a height that will cover the rest of the bands when the rectangle is copied and dragged to the next band as the boxes will end up in a straight line.
20. Some of the bands will not generate a peak that is above the background.
21. The total protein area value (supernatant + pellet) should be equal or very similar to the value of the supernatant of the corresponding buffer control.

References

1. Peter BJ, Kent HM, Mills IG, Vallis Y, Butler PJ, Evans PR, McMahon HT (2004) BAR domains as sensors of membrane curvature: the amphiphysin BAR structure. Science 303:495–499
2. Daumke O, Lundmark R, Vallis Y, Martens S, Butler PJ, McMahon HT (2007) Architectural and mechanistic insights into an EHD ATPase involved in membrane remodelling. Nature 449:923–927
3. Hill MM, Bastiani M, Luetterforst R, Kirkham M, Kirkham A, Nixon SJ, Walser P, Abankwa D, Oorschot VM, Martin S, Hancock JF, Parton RG (2008) PTRF-Cavin, a conserved cytoplasmic protein required for caveola formation and function. Cell 132:113–124
4. Mohan J, Moren B, Larsson E, Holst MR, Lundmark R (2015) Cavin3 interacts with cavin1 and caveolin1 to increase surface dynamics of caveolae. J Cell Sci 128:979–991
5. Kovtun O, Tillu VA, Jung W, Leneva N, Ariotti N, Chaudhary N, Mandyam RA, Ferguson C, Morgan GP, Johnston WA, Harrop SJ, Alexandrov K, Parton RG, Collins BM (2014) Structural insights into the organization of the cavin membrane coat complex. Dev Cell 31:405–419
6. Zhao H, Lappalainen P (2012) A simple guide to biochemical approaches for analyzing protein-lipid interactions. Mol Biol Cell 23:2823–2830
7. Pylypenko O, Lundmark R, Rasmuson E, Carlsson SR, Rak A (2007) The PX-BAR membrane-remodeling unit of sorting nexin 9. EMBO J 26:4788–4800
8. Lundmark R, Doherty GJ, Vallis Y, Peter BJ, McMahon HT (2008) Arf family GTP loading is activated by, and generates, positive membrane curvature. Biochem J 414:189–194
9. Lundmark R, Doherty GJ, Howes MT, Cortese K, Vallis Y, Parton RG, McMahon HT (2008) The GTPase-activating protein GRAF1 regulates the CLIC/GEEC endocytic pathway. Curr Biol 18:1802–1808
10. Hoernke M, Mohan J, Larsson E, Blomberg J, Kahra D, Westenhoff S, Schwieger C, Lundmark R (2017) EHD2 restrains dynamics of caveolae by an ATP-dependent, membrane-bound, open conformation. Proc Natl Acad Sci U S A 114(22):E4360–E4369. https://doi.org/10.1073/pnas.1614066114

Chapter 12

Liposome Binding Assay to Characterize the Structure and Function of Cavin Proteins

Miriam Stoeber

Abstract

Protein–protein and protein–lipid interactions play important roles in the assembly of protein coats that regulate membrane organization, signaling, and trafficking in eukaryotic cells. Caveolae are plasma membrane invaginations that are formed by a protein coat consisting of caveolin and cavin protein complexes. The biochemical and structural principles of membrane binding by coat components can be studied through in vitro reconstitution of purified proteins and lipid vesicles. In this chapter, we describe a method to isolate peripheral cavin coat complexes and to subsequently bind purified cavin to chemically defined liposomes. The cavin proteoliposomes can be further analyzed to gain insights into lipid binding specificity, membrane-remodeling properties, and structural characteristics of the cavin family members.

Key words Caveolae, Cavin, Coat protein, Liposome binding, Liposome flotation, Reconstitution, Proteoliposome

1 Introduction

Oligomers of membrane-embedded caveolins and peripherally attached cavins form the protein coat of caveolae. Each caveolar coat contains a defined number of both of these proteins that form independent assemblies with sedimentation coefficients of 70 S and 60 S, respectively [1–4]. While caveolins are integral membrane proteins with a hydrophobic hairpin loop embedded in the lipid bilayer, cavins are soluble proteins that are recruited from the cytosol to the caveolar membrane domain defined by caveolin oligomers [1, 5–8]. Cavins are essential for stabilizing caveolae and promoting the invaginated shape. Since cavins show little or no direct interaction with caveolin, it is likely that cavins associate with caveolae mainly via binding to membrane lipids [5, 7, 9, 10]. Lipid species that are enriched in caveolae include cholesterol, glycosphingolipids, phosphatidylserine (PS), and certain phosphoinositides (i.e. phosphatidylinositol (4,5)-bisphosphate, PIP_2) [11–13]. All cavin family members have been shown to bind negatively

Cedric M. Blouin (ed.), *Caveolae: Methods and Protocols*, Methods in Molecular Biology, vol. 2169,
https://doi.org/10.1007/978-1-0716-0732-9_12,

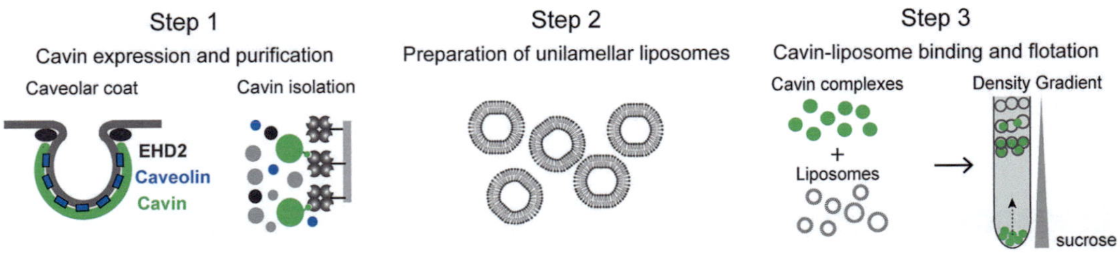

Fig. 1 Workflow of the liposome binding assay to characterize cavin protein structure and functions. Three separate steps are necessary to obtain cavin proteoliposomes (*see* Subheading 3)

charged PS, which suggests that they are recruited to caveolae by electrostatic interactions with lipid head groups [7, 10, 14, 15]. In order to study membrane binding and remodeling characteristics of cavins as well as the structure of the peripheral cavin coat, we developed a liposome binding assay for purified cavins. The protocol (Fig. 1) described here was used for isolation of Strep-tagged wild-type Cavin1 60S complexes and smaller oligomers formed by mutants of Cavin1 [10]. Liposome binding and flotation were then performed using PS-containing unilamellar vesicles with a size similar to caveolae (80 nm). The protocol can be easily adapted to other members of the cavin family (Cavin2, 3, and 4)—or de facto any soluble candidate lipid-binding protein—and to liposomes with different lipid composition and size. The resulting proteoliposomes are well suited for downstream studies that investigate the lipid binding and structural characteristics of cavins, which are critical to understanding the cellular functions of cavins and caveolae.

2 Materials

2.1 Purification of Cavin Complexes

1. Eukaryotic cell line (e.g. HEK293) expressing recombinant StrepIII-tagged Cavin1.
2. StrepTactin Sepharose suspension and empty gravity-flow columns.
3. TNE buffer: 20 mM Tris–HCl pH 7.4, 250 mM NaCl, 5 mM EDTA (*see* **Note 1**).
4. Lysis buffer: TNE buffer, 1% Triton X-100, protease inhibitors. Prepare fresh and put on ice.
5. Elution buffer: TNE buffer, 2.5 mM desthiobiotin. Prepare fresh.

2.2 Preparation of Liposomes

1. Chloroform-solubilized lipids (stock 10 mg/mL, Avanti Polar Lipids): 1,2-di-(9Z-octadecenoyl)-sn-glycero-3-phosphocholine (DOPC), DO-phosphoethanolamine (DOPE), DO-phospho-L-serine (DOPS), and cholesterol (*see* **Note 2**).

2. Argon or nitrogen stream to evaporate chloroform solvent of lipid mixture.
3. Liquid nitrogen for freeze-thaw cycles.
4. Avanti Mini Extruder, 80 nm polycarbonate membrane.

2.3 Liposome Flotation

1. Ultracentrifuge with swing-bucket rotor (e.g. SW55) and ultra-clear tubes.
2. Light and heavy sucrose density media: 5% and 35% sucrose in TNE buffer (*see* **Note 3**).
3. Gradient-forming instrument (e.g. Gradient Master with tilted tube rotation) to produce linear gradients.

3 Methods

3.1 One-Step Purification of Cavin Complexes

1. Express Strep-tagged Cavin1: Transfect two T175 flasks containing adherent, 70% confluent HEK293 cells with recombinant StrepIII-tagged Cavin1 plasmid. Allow protein expression for 24–48 h (*see* **Notes 4–6**).
2. Detach and wash cells: Take flasks out of the cell culture incubator and wash cells two times with 10 mL cold PBS. Add 3 mL of cold PBS to each flask and detach cells with a cell scraper. Transfer cell suspension to a 50-mL Falcon tube and spin down cells (500 × *g*, 5 min). Discard PBS. Resuspend cell pellet in 30 mL cold PBS and spin down again (500 × *g*, 5 min). Discard PBS.
3. Prepare cleared lysate: Add 2 mL ice-cold lysis buffer to cells and resuspend by pipetting up and down with P1000 tip. Pass cell lysate ten times through 25G syringe to mechanically break remaining cells. Keep lysate on ice for 20 min. Centrifuge lysate to pellet nuclei (1000 × *g*, 5 min at 4 °C). Transfer postnuclear supernatant into two new 1.5 mL Eppendorf tubes and spin lysates to pellet cell debris (15,000 × *g*, 15 min at 4 °C).
4. Prepare StrepTactin Sepharose beads: Use 200 μL of 50% StrepTactin Sepharose suspension resin (results in column volume of 100 μL). Apply beads to empty gravity-flow column. Wash and equilibrate beads by passing three times 500 μL TNE buffer.
5. Binding and washing: Add cleared lysate in 500 μL steps to StrepTactin Sepharose beads. Let lysate run through the column by gravity flow. Reload lysate one more time. Wash column three times with 500 μL lysis buffer. Then wash column three times with 500 μL TNE buffer (no detergent).
6. Elution: Sequentially elute Cavin1 from the column by passing five times 100 μL elution buffer. Collect all five elutions and

keep purified cavin samples at 4 °C until use in subsequent steps (*see* **Note 7**).

7. Evaluate purified material: Measure protein concentration, determine purity (e.g. by SDS-Coomassie or mass-spec analysis), and test integrity of cavin complex (e.g. by sucrose velocity gradient analysis or Small-angle X-ray scattering) (*see* **Note 8**).

3.2 Preparation of Unilamellar Lipid Vesicles

1. Prepare lipid mix: In a round bottom glass tube, combine a total of 2 nanomoles chloroform-solubilized lipids (approximately 1.6 mg) using the desired molar fraction of each lipid species (e.g. DOPC 25 mol%, DOPE 20 mol%, DOPS 5 mol%, and cholesterol 50 mol%).
2. Evaporate chloroform: Blow-dry the lipid mix using an argon or nitrogen gas stream for 15 min to form a lipid cake at the glass tube bottom.
3. Hydrate lipids: When the solvent is dried, immediately add 1.6 mL TNE buffer to lipids to obtain a lipid solution at 1.25 mM. Vortex thoroughly for 2–3 min until the solution looks homogenous and cloudy (*see* **Note 9**).
4. Prepare unilamellar liposomes: Transfer 250 μL of the lipid solution into a 1.5-mL Eppendorf tube and perform freeze-thaw cycles: Place the tube for 3 min in liquid nitrogen, then place the tube for 3 min at 60 °C, repeat ten times in total.
5. Extrude liposomes at defined size: Assemble the polycarbonate membrane (80 nm pore size) inside the extruder, pre-wet the Teflon piece, the two pieces of membrane support in the center of the Teflon piece, and syringes with TNE buffer. Fill 250 μL of the lipid solution into the loading syringe and extrude 11 times. Eject extruded lipids from the acceptor syringe to a new 1.5 mL tube. The lipid solution will look clear at this step. Store lipids at 4 °C until use (*see* **Notes 10–12**).

3.3 Cavin Binding to Liposomes and Proteoliposome Flotation

1. Incubation of Cavin1 with liposomes: Use 200 μL of freshly prepared unilamellar liposome solution (approx. 200 μg lipids) and combine with 25–75 μg purified Cavin1 (protein:lipid ratio between 1:10 and 1:2), incubate liposomes and Cavin1 for 1 h at room temperature (*see* **Note 13**).
2. Preparation of density gradients with a "Gradient Master": Prepare gradient solutions (5% and 35% sucrose) in TNE buffer. Prepare 3 mL of each solution per gradient. Insert ultra-clear centrifugation tube into the "Marker Block" and use a pen to mark the half-filling position of the tube with the help of the lower "Marker Block" line. Pour 35% sucrose solution to half-filling position, carefully overlay with 5% sucrose solution. Use long caps to seal the tubes. Insert tubes into the tube holder and place it on the tilting plate of

"Gradient Master." Prepare linear 5–35% sucrose gradients using the following parameters: angle: 86.5°, time: 35 s, speed: 28. After rotation, place the tube holder back on the bench, remove any access sucrose solution that may have leaked into the caps, and carefully remove caps from gradients (*see* **Notes 14** and **15**).

3. Load sample: Adjust sucrose concentration of reconstituted sample to 40% by adding 70% sucrose stock solution in TNE buffer. Load 750 μL of sucrose-adjusted sample in a syringe with a long, 50-mm needle. Carefully insert the needle into the gradient all the way to the bottom of the centrifugation tube. Slowly eject the sample so that it remains at the bottom and pushes gradient media up.
4. Run density gradient and fractionate: Centrifuge at 4 °C, 40.000 rpm for 16 h using the SW55 swing bucket rotor. After the run is completed, carefully transfer the tubes back to the tube holder. Fractionate the gradient into 14 fractions of 360 μL with increasing density from the top to the bottom of gradient. Carefully touch the surface of the gradient liquid with the pipett tip and gradually lower the tip while pipetting off the fractions.
5. Evaluate flotation result: Measure the protein amount in each fraction (e.g. by SDS-PAGE/Western blotting/Coomassie staining or dot blot). Unbound protein will remain at the bottom of the gradient, liposome-associated complexes will float to the top fractions of the gradient (Fig. 1) (*see* **Notes 16–18**).

4 Notes

1. TNE buffer can be prepared as 5× stock solution (100 mM Tris–HCl pH 7.4, 1.25 M NaCl, 25 mM 25 mM EDTA) and stored for several weeks at room temperature.
2. The lipid composition of the vesicles can be changed according to the desired experimental condition, e.g. different phosphoinositides can be added. We recommend staying close to the physiological steady-state composition of cellular membranes [16].
3. A 60–70% sucrose stock solution in TNE buffer can be prepared and stored for several days. Warm up solution to 37 °C to improve sucrose solubility.
4. Different cell lines are suitable for expression of cavin proteins, e.g. HEK293, HeLa, and CV-1 cells. To avoid transient transfection of cells, we have generated a Flp-In-293 cell line

expressing a tetracycline-inducible, N-terminally tagged StrepIII-Cavin1 construct.

5. Thus far, we have not been able to produce cavin 60S complexes in bacteria using various expression vectors and cavin co-expression systems (unpublished results). Therefore, we recommend using mammalian cells for expression and purification of full-length cavins. However, it is possible to produce smaller complexes in bacteria, e.g. formed by truncated cavin proteins, as previously described [15, 17]. These can also be used is subsequent reconstitution steps.
6. The small-scale purification protocol described here can be easily scaled up using more cell material and adapting the amounts of Strep-Tactin Sepharose beads and buffers used in each step. We scaled up the purification by 15 times by growing HEK293 cells in roller bottles instead of T175 flasks.
7. The protein concentration of our elution samples is usually between 0.25 and 0.4 mg/mL. We tried to concentrate purified Cavin1 beyond 0.5 mg/mL using ultrafiltration centrifugal devices. However, size-exclusion chromatography and cryo-EM analysis of the concentrated cavin sample revealed cavin aggregates of large sizes (unpublished results) and thus, concentrating the eluted samples is not advised.
8. We recommend storing purified 60S cavin complexes in TNE buffer at 4 °C after purification for no longer than 4–5 days. Using sucrose gradient analysis, we found that 60S cavin complexes fully dissociate within a week of purification.
9. At this step, the cloudy lipid solution contains multilamellar liposomes of various sizes that can be stored for several months at −20 °C. We prepared several 250 μL aliquots that we thawed and used at later points in the subsequent steps.
10. The lipid solution has to be kept above the phase transition temperature of the lipids during hydration and extrusion. It may be necessary to heat lipid mix if the transition temperature is higher than room temperature (which is not the case for the indicated lipid composition).
11. The size of liposomes can easily be changed using different polycarbonate membranes with pore sizes ranging from 30 to 200 nm. Even larger membrane pore sizes will result in multilamellar liposomal suspensions. The liposome preparation should be examined regarding vesicle sizes and shapes, e.g. by electron microscopy.

12. The small, unilamellar vesicles can be stored in aqueous media at 4 °C. However, the liposome solution is not stable for longer than 2–3 days.
13. We recommend varying the ratio of cavin:liposomes to determine the protein binding capacity of cavins to chemically distinct vesicles. Unbound protein will remain at the bottom of the density gradient.
14. If the presence of sucrose interferes with downstream applications, sucrose can be exchanged to other density gradient media. We have previously used the nonionic, iodinated media Nycodenz and prepared 30–60% (vol/vol) Nycodenz gradients.
15. Instead of linear density gradients, flotation analysis can also be performed using step gradients. We have successfully used step gradients with three layers (35%, 20%, and 0% sucrose). Here, centrifugation was performed at 50.000 rpm for 2.5 h at 4 °C and protein-coated liposomes could be collected at the 20-0% sucrose interface (unpublished results). The advantage of linear density gradients and equilibrium centrifugation is the clear separation of unbound liposomes (lightest, top fraction) from protein-bound liposomes (heavier fractions).
16. The flotation pattern of proteoliposomes can be evaluated visually or with a fluorescence imager when small amounts of fluorescently conjugated phospholipids are added to the lipid mix. For example, we have included 1 mol% Rhodamine-PE that enabled us to see a red "liposome band" after centrifugation.
17. Spreading of proteoliposomes over many fractions of the linear density gradient indicates the presence of vesicles with varying protein:lipid ratio. This can be caused by a heterogenous liposome preparation (e.g. containing liposomes of different sizes or multilamellar vesicles) or by a potential membrane-remodeling activity of the cavin family member under investigation.
18. Proteoliposome flotation may be omitted if downstream applications do not require the separation of bound from unbound vesicles and proteins (e.g. EM analysis).

Acknowledgments

This work was supported by the European Molecular Biology Organisation (EMBO) Short-Term Fellowship 36-2013. M.S. is supported by the SNSF Eccellenza Professorial Fellowship (PCEFP3_181282).

References

1. Hayer A, Stoeber M, Bissig C, Helenius A (2010) Biogenesis of caveolae: stepwise assembly of large caveolin and cavin complexes. Traffic 11:361–382
2. Bastiani M, Liu L, Hill MM et al (2009) MURC/Cavin-4 and cavin family members form tissue-specific caveolar complexes. J Cell Biol 185:1259–1273
3. Ludwig A, Howard G, Mendoza-Topaz C et al (2013) Molecular composition and ultrastructure of the caveolar coat complex. PLoS Biol 11:e1001640–e1001620
4. Gambin Y, Ariotti N, McMahon K-A et al (2014) Single-molecule analysis reveals self-assembly and nanoscale segregation of two distinct cavin subcomplexes on caveolae. Elife 3:3831–3818
5. Kovtun O, Tillu VA, Ariotti N et al (2015) Cavin family proteins and the assembly of caveolae. J Cell Sci 128:1269–1278
6. Lamaze C, Tardif N, Dewulf M et al (2017) The caveolae dress code: structure and signaling. Curr Opin Cell Biol 47:117–125
7. Hill MM, Bastiani M, Luetterforst R et al (2008) PTRF-cavin, a conserved cytoplasmic protein required for caveola formation and function. Cell 132:113–124
8. Parton RG, Hanzal-Bayer M, Hancock JF (2006) Biogenesis of caveolae: a structural model for caveolin-induced domain formation. J Cell Sci 119:787–796
9. Hirama T, Das R, Yang Y et al (2017) Phosphatidylserine dictates the assembly and dynamics of caveolae in the plasma membrane. J Biol Chem 292:14292–14307
10. Stoeber M, Schellenberger P, Siebert CA et al (2016) Model for the architecture of caveolae based on a flexible, net-like assembly of Cavin1 and Caveolin discs. Proc Natl Acad Sci U S A 113:E8069–E8078
11. Fairn GD, Schieber NL, Ariotti N et al (2011) High-resolution mapping reveals topologically distinct cellular pools of phosphatidylserine. J Cell Biol 194:257–275
12. Wanaski SP, Ng BK, Glaser M (2003) Caveolin scaffolding region and the membrane binding region of src form lateral membrane domains. Biochemistry 42:42–56
13. Ortegren U, Karlsson M, Blazic N et al (2004) Lipids and glycosphingolipids in caveolae and surrounding plasma membrane of primary rat adipocytes. Eur J Biochem 271:2028–2036
14. Hansen CG, Nichols BJ (2010) Exploring the caves: cavins, caveolins and caveolae. Trends Cell Biol 20:177–186
15. Kovtun O, Tillu VA, Jung W et al (2014) Structural insights into the organization of the cavin membrane coat complex. Dev Cell 31:405–419
16. van Meer G, Voelker DR, Feigenson GW (2008) Membrane lipids: where they are and how they behave. Nat Rev Mol Cell Biol 9:112–124
17. Hansen CG, Bright NA, Howard G, Nichols BJ (2009) SDPR induces membrane curvature and functions in the formation of caveolae. Nat Cell Biol 11:807–814

Chapter 13

Preparation of Caveolin-1 for NMR Spectroscopy Experiments

Sarah M. Plucinsky, Jeffrey A. Julien, and Kerney Jebrell Glover

Abstract

Caveolin-1 is a 20.5 kDa integral membrane protein that is involved in a myriad of cellular processes including signal transduction, relieving mechano-stresses on the cell, endocytosis, and most importantly caveolae formation. As a consequence, there is intense interest in characterizing caveolin-1 structurally. Out of the many available structural techniques, nuclear magnetic resonance (NMR) spectroscopy is particularly well suited to investigations on integral membrane proteins like caveolin-1 that have significant unstructured regions and unusual topologies. However, the technique requires relatively large amounts of protein (i.e. concentrations in the 0.5–5 mM range), and obtaining these amounts can be difficult especially for highly hydrophobic membrane proteins such as caveolin-1. Herein, we describe a robust protocol for the preparation of caveolin-1 for structural studies using NMR.

Key words Caveolin, NMR spectroscopy, Inclusion bodies, Membrane protein purification, HPLC, Cyanogen bromide cleavage

1 Introduction

Nuclear magnetic resonance (NMR) spectroscopy has emerged as a powerful tool for protein structure elucidation. It is especially attractive for membrane proteins because, unlike techniques such as X-ray crystallography where lipids and detergents can present significant challenges, it is widely compatible with many membrane mimics (i.e. micelles, bicelles, and nanodiscs) [1–4]. In addition, NMR spectroscopy allows for structure determination in solution, which may provide key information concerning the dynamic state of the protein [5]. Another advantage of NMR spectroscopy relative to other spectroscopic methods such as circular dichroism spectroscopy is that it provides residue-specific structural information.

NMR spectroscopy requires millimolar quantities of protein which can present a significant challenge for membrane proteins that typically have lower overexpression levels and more

Cedric M. Blouin (ed.), *Caveolae: Methods and Protocols*, Methods in Molecular Biology, vol. 2169,
https://doi.org/10.1007/978-1-0716-0732-9_13,

sophisticated purification protocols due to their hydrophobicity. Caveolin-1 is one such example. It is a 20.5 kDa highly insoluble integral membrane protein that plays a preeminent role in the biogenesis of caveolae, which are cell-surface invaginations that punctuate the plasma membrane of many mammalian cell types [6]. Additionally, caveolin-1 has been shown to be intimately related to several functions of caveolae including endocytosis, signal transduction, and mechano-protection [7–10]. The hallmark structural feature of caveolin is an uncommonly long hydrophobic span that is thought to form an intramembrane turn within the anhydrous core of the bilayer, which has sparked intense interest in caveolin structure determination. Herein, we describe a robust protocol to obtain homogeneously purified caveolin-1 for structural studies using solution-state NMR, and note that the methodology presented has allowed for the first structural characterization of a functional construct of caveolin-1 [11].

2 Materials

All solutions were prepared using deionized water. $^{15}NH_4Cl$, ^{13}C-glucose, and D_2O were purchased from Cambridge Isotopes (Tewksbury, MA). 1-Myristoyl-2-hydroxy-sn-glycero-3-phospho-(1′-rac-glycerol) (sodium salt) (LMPG) was purchased from Avanti Polar lipids (Alabaster, Alabama). All other reagents were of standard laboratory grade.

2.1 Media for ^{15}N-Labeled Caveolin-1

1. Trace metals (100 mL): 36 mL H_2O (autoclaved), 50 mL 0.1 M $FeCl_3$ in 0.12 M HCl, 2 mL 1 M $CaCl_2$ (autoclaved), 1 mL 1 M $MnCl_2{\cdot}4H_2O$ (autoclaved), 1 mL 1 M $ZnSO_4{\cdot}7H_2O$ (autoclaved), 1 mL 0.2 M $CoCl_2{\cdot}6H_2O$ (autoclaved), 2 mL 0.1 M $CuCl_2{\cdot}2H_2O$ (autoclaved), 1 mL 0.2 M $NiCl_2{\cdot}6H_2O$ (autoclaved), 2 mL 0.1 M $Na_2MoO_4{\cdot}2H_2O$ (autoclaved), 2 mL 0.1 M $Na_2SeO_3{\cdot}5H_2O$ (autoclaved), 2 mL 0.1 M H_3BO_3 (autoclaved). The above recipe is based on the autoinduction methodology developed by Studier [12].
2. 50× Salts (1 L): Add 178 g Na_2HPO_4, 170 g KH_2PO_4, 134 g NH_4Cl, and 36 g Na_2SO_4 to 800 mL of water. Filter (0.2 μm) and autoclave. The above recipe is based on the autoinduction methodology developed by Studier [12].
3. 50× Sugars (1 L): Add 250 g glycerol, 25 g D-glucose, 100 g α-D-lactose monohydrate to 730 mL of water. Sterile filter (0.2 μm). The above recipe is based on the autoinduction methodology developed by Studier [12].
4. MDG media (5 mL): 10 μL 1 M $MgSO_4$ (autoclaved), 1 μL trace metals, 62.5 μL 40% (w/v) D-glucose (sterile filtered through 0.2 μm), 50 μL 25% (w/v) aspartate (autoclaved),

100 μL 50× Salts, 4.8 mL water (autoclaved), 5 μL 30 mg/mL kanamycin (sterile filtered through 0.2 μm). The above recipe is based on the autoinduction methodology developed by Studier [12].

5. N-5052 media (1 L): Add 7.10 g Na_2HPO_4, 6.80 g KH_2PO_4, 2.73 g $^{15}NH_4Cl$, and 0.710 g Na_2SO_4 to 978 mL of water and autoclave. After the solution has cooled, add 2 mL 1 M $MgSO_4$ (autoclaved), 200 μL trace metals, 20 mL 50× Sugars. The above recipe is based on the autoinduction methodology developed by Studier [12].
6. 0.9% (w/v) saline (2 L): Add 18 g NaCl to 2 L of deionized water. Store at 4 °C.

2.2 Media for ^{13}C,^{15}N-Labeled Caveolin-1

1. LB media (1 L): Add 10 g tryptone, 10 g sodium chloride, and 5 g yeast extract to 800 mL of water. Adjust the pH to 7.0 with NaOH, bring the volume to 1 L using a graduated cylinder, and autoclave.
2. 5× M9 Salts (1 L): Add 15 g KH_2PO_4, 34 g Na_2HPO_4, 2.5 g NaCl, 5 g NH_4Cl, bring the volume to 1 L using a graduated cylinder, and autoclave.
3. 5× M9 Salts with ^{15}N (1 L): Add 15 g KH_2PO_4, 34 g Na_2HPO_4, 2.5 g NaCl, 5 g $^{15}NH_4Cl$, bring the volume to 1 L using a graduated cylinder, and autoclave.
4. M9 minimal media (1 L): Add 778 mL water to a 6 L Erlenmeyer flask and autoclave. When cool, add 200 mL of 5× M9 Salts, 20 mL 0.2 g/mL D-glucose (sterile filtered through 0.2 μm), 2 mL 1 M MgSO4 (autoclaved), and 0.1 mL 1 M $CaCl_2$ (autoclaved).
5. M9 minimal media with ^{13}C and ^{15}N (1 L): Add 778 mL water to a 6 L Erlenmeyer flask and autoclave. When cool, add 200 mL 5× M9 Salts with ^{15}N, 20 mL 0.2 g/mL ^{13}C D-glucose (sterile filtered through 0.2 μm), 2 mL 1 M $MgSO_4$ (autoclaved), and 0.1 mL 1 M $CaCl_2$ (autoclaved).
6. 1 M IPTG (Isopropyl-β-D-thiogalactoside) (20 mL): Add 4.76 g IPTG and bring the volume to 20 mL with a graduated cylinder (sterile filtered through 0.2 μm). Store in 1 mL aliquots at −20 °C.

2.3 Media for ^{13}C,^{15}N,^{2}H-Labeled Caveolin-1

1. M9 minimal media with ^{13}C, ^{15}N, and ^{2}H (1 L): Add 778 mL D_2O to a 6 L Erlenmeyer flask and autoclave. When cool, add 200 mL 5× M9 Salts with ^{15}N, 20 mL 0.2 g/mL ^{13}C D-glucose (sterile filtered through 0.2 μm), 2 mL 1 M $MgSO_4$ (autoclaved), and 0.1 mL 1 M $CaCl_2$ (autoclaved).

2.4 Buffers for Inclusion Body Isolation

1. 1 M Tris pH 8.0 (1 L): Add 121.14 g Tris to 800 mL water. Adjust the pH to 8.0 using concentrated HCl and bring the total volume of the solution to 1 L using a graduated cylinder.
2. Sucrose buffer (1 L): Add 200 g sucrose, 20 mL 1 M Tris pH 8.0, 1.4 mL of neat 2-mercaptoethanol (14.3 M), and bring the total volume of the solution to 1 L using a graduated cylinder.
3. Detergent buffer (1 L): Add 10 mL of Triton™ X-100, 20 mL Tris pH 8.0, 1.4 mL of neat 2-mercaptoethanol (14.3 M), and bring the total volume of the solution to 1 L using a graduated cylinder.

2.5 Buffers for HPLC Purification

1. HPLC buffer A (1 L): 80% (v/v) water, 20% (v/v) acetic acid. Mix 800 mL of water and 200 mL of glacial acetic acid. Filter (0.2 μm).
2. HPLC buffer B (1 L): 80% (v/v) 1-butanol, 20% (v/v) acetic acid. Mix 800 mL of 1-butanol and 200 mL of glacial acetic acid. Filter (0.2 μm).

2.6 Buffers for Preparation of NMR Sample

1. NMR sample buffer: 100 mM LMPG (1-myristoyl-2-hydroxy-sn-glycero-3-phospho-(1′-rac-glycerol) (sodium salt)), 20 mM phosphate pH 7.0, 100 mM NaCl and 10% (v/v) D_2O. Add 0.478 g LMPG, 0.028 g NaH_2PO_4, 0.1 g NaCl, 1 mL of D_2O, and 5 mL of H_2O. Adjust the pH to 7.0 using a concentrated solution of NaOH and bring the volume to 10 mL using a graduated cylinder.

3 Methods

3.1 Preparation of Construct

1. Residues 62–178 of human caveolin-1 was cloned into a pET-24a vector as a C-terminal fusion to the TrpLE (Trp leader) protein.
   (a) TrpLE sequence:
   KAIFVLKGSLDRDLDSRIELELRTDHKELSEHLLLV
   DLARNDLARIATPGSRYVADLTKVDRYSYVLHLVSR
   VVGELRHDLDALHAYRAALNLGTLSGAPKVRAGG
   LVPRGSM
   (b) Caveolin-1 (62–178) sequence:
   DVVKIDFEDVIAEPEGTHSFDGIWKASFTTFTVTKY
   WFYRLLSALFGIPLALIWGIYFAILSFLHIWAVVPSIK
   SFLIEIQSISRVYSIYVHTVSDPLFEAVGKIFSNV
   RINLQKEI
2. Mutations: C133S, C143S, C156S, and M111L (*see* **Notes 1–4**).

3.2 Expression of ^{15}N-Labeled Caveolin-1 (62–178)

The overexpression procedure is based on the autoinduction methodology developed by Studier [12].

1. Put 5 mL of sterile MDG media into a sterile glass test tube (17 mm × 100 mm).
2. Inoculate the MDG with BL-21 cells that have been transformed with the pET(TrpLE-M-Cav1$_{62–178}$)-24a vector.
3. Shake at 250 rpm for 20 h at 37 °C.
4. Add 1 mL of the 20 h culture to 1 L of sterile N-5052 media in a 6 L Erlenmeyer flask.
5. Shake at 250 rpm for 18 h at 37 °C.
6. Harvest cells by centrifugation (30 min, 5000 x g, 4 °C).
7. Wash cell pellet with 200 mL of 0.9% (w/v) saline and recentrifuge (30 min, 5000 x *g*, 4 °C). Pellets can be stored at −80 °C indefinitely.

3.3 Expression of ^{15}N and ^{13}C-Labeled Caveolin-1 (62–178)

The overexpression procedure is based on the methodology developed by Marley et al. [14].

1. Put 5 mL of sterile MDG media into a sterile glass test tube (17 mm × 100 mm).
2. Inoculate the MDG with BL-21 cells that have been transformed with the pET(TrpLE-M-Cav1$_{62–178}$)-24a vector.
3. Shake at 250 rpm for 20 h at 37 °C.
4. Add 1 mL of the 20 h culture to 1 L of sterile LB media in a 6 L Erlenmeyer flask.
5. Grow to an OD_{600} of 0.8 and harvest cells by centrifugation (30 min, 5000 x *g*, 25 °C) (*see* **Note 5**).
6. Wash cell pellet with 500 mL of M9 minimal media and recentrifuge (30 min, 5000 x *g*, 25 °C) (*see* **Note 5**).
7. Resuspend pellet in 500 mL M9 minimal media with ^{13}C and ^{15}N and shake for 2 h (250 rpm, 37 °C).
8. Induce protein expression by adding 500 μL of 1 M IPTG and continue to shake for 8 h (250 rpm, 37 °C).
9. Harvest cells by centrifugation (30 min, 5000 x *g*, 4 °C).
10. Wash pellet with 200 mL of 0.9% (w/v) saline and recentrifuge (30 min, 5000 x *g*, 4 °C). Pellets can be stored at −80 °C indefinitely.

3.4 Expression of ^{15}N, ^{13}C and ^{2}H-Labeled Caveolin-1 (62–178)

The overexpression procedure is based on the methodology developed by Marley et al. [14].

1. Put 5 mL of sterile MDG media into a sterile glass test tube (17 mm × 100 mm).

2. Inoculate the MDG with BL-21 cells that have been transformed with the pET(TrpLE-M-Cav1$_{62–178}$)-24a vector.
3. Shake at 250 rpm for 20 h at 37 °C.
4. Add 1 mL of the 20 h culture to 1 L of sterile LB media in a 6 L Erlenmeyer flask.
5. Grow to an OD_{600} of 0.8 and harvest cells by centrifugation at (30 min, 5000 x *g*, 25 °C) (*see* **Note 5**).
6. Wash cell pellet with 500 mL M9 minimal media and recentrifuge (30 min, 5000 x *g*, 25 °C) (*see* **Note 5**).
7. Resuspend pellet in 500 mL M9 minimal media with ^{13}C, ^{15}N, and ^{2}H and shake for 2 h (250 rpm, 37 °C).
8. Induce protein expression by adding 500 μL of 1 M IPTG and continue to shake for 8 h (250 rpm, 37 °C).
9. Harvest cells by centrifugation (30 min, 5000 x *g*, 4 °C).
10. Wash pellet with 200 mL of 0.9% (w/v) saline and recentrifuge (30 min, 5000 x *g*, 4 °C). Pellets can be stored at −80 °C indefinitely.

3.5 Purification of Caveolin-1 (62–178)

The purification procedure is based on the method described in Diefenderfer et al. [15].

1. Resuspend the cell pellet in 200 mL of Sucrose buffer and lyse cells by sonication with stirring (Branson Sonifier™ 450, duty cycle 4, power 5) in 10-min intervals for 30 min total (10 min on, 5 min off).
2. Remove 100 μL of the suspension and save for analysis by SDS-PAGE (sodium dodecyl sulfate polyacrylamide gel electrophoresis). This aliquot is called the "pre-sucrose" sample.
3. Centrifuge suspension for 2 h (27,000 x *g*, 15 °C).
4. Remove 100 μL of the supernatant and save for analysis by SDS-PAGE. This aliquot is called the "post-sucrose" sample.
5. Pour off the supernatant (*see* **Notes 8** and **9**).
6. Resuspend pellet in 200 mL of Detergent buffer and sonicate with stirring (Branson sonifier 450, duty cycle 4, power 5) for 10-min intervals for 30 min total (10 min on, 5 min off) (*see* **Note 6**).
7. Remove 100 μL of the suspension and save for analysis by SDS-PAGE. This aliquot is called the "pre-detergent" sample.
8. Centrifuge suspension for 2 h (27,000 x *g*, 15 °C) (*see* **Note 7**).
9. Remove 100 μL of the supernatant and save for analysis by SDS-PAGE. This aliquot is called the "post-detergent" sample.
10. Pour off the supernatant (*see* **Notes 8** and **9**).

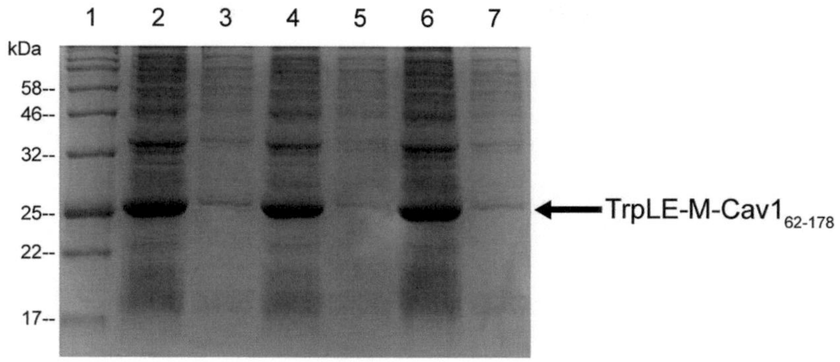

Fig. 1 Coomassie-stained 12% polyacrylamide gel. Lane 1 molecular weight ladder. Lanes 2, 4, 6 pre-Sucrose, -Detergent, and -IPA samples. Lanes 3, 5, 7 post-Sucrose, -Detergent, and -IPA samples. 20 μL was loaded for all samples at 50% dilution with 2× reducing SDS-PAGE loading buffer

11. Resuspend pellet in 80 mL of 33 mM Tris pH 8.0 (2.64 mL 1 M Tris pH 8.0, 77.4 mL water) and sonicate for 10 min (Branson sonifier 450, duty cycle 4, power 5).
12. After sonication, add 120 mL of IPA (isopropyl alcohol) and mix.
13. Remove 100 μL of the suspension and save for analysis by SDS-PAGE. This aliquot is called the "pre-IPA" sample.
14. Centrifuge suspension for 2 h (27,500 x *g*, 15 °C).
15. Remove 100 μL of the supernatant and save for analysis by SDS-PAGE. This aliquot is called the "post-IPA" sample.
16. Pour off the supernatant (*see* **Notes 8** and **9**).
17. Upon completion of the washes, the "pre" and "post" aliquots are analyzed by SDS-PAGE (Fig. 1).
18. After confirmation by SDS-PAGE, the pellet was resuspended in 30 mL of 88% formic acid (*see* **Note 10**).
19. Add 0.2 g of cyanogen bromide and bubble with N_2 for 5 min. Rotate the sample for 18 h at room temperature.
20. The formic acid was removed using a centrifugal concentrator (Thermo Savant™ SPD121P) (*see* **Note 11**).
21. The film was dissolved into 9 mL of 88% formic acid and filtered through 0.2 μm (*see* **Note 12**).
22. The mixture was injected on a reversed-phase C4 column HPLC (high-performance liquid chromatography) utilizing a 1% gradient from 5 to 100% HPLC buffer B at a flow rate of 10 mL/min (Fig. 2).
23. The starred peak was collected and dried under vacuum into a film (Fig. 2).
24. The film is then taken up into 5 mL of Hexafluoroisopropanol (HFIP), flash frozen with liquid N_2, and lyophilized.

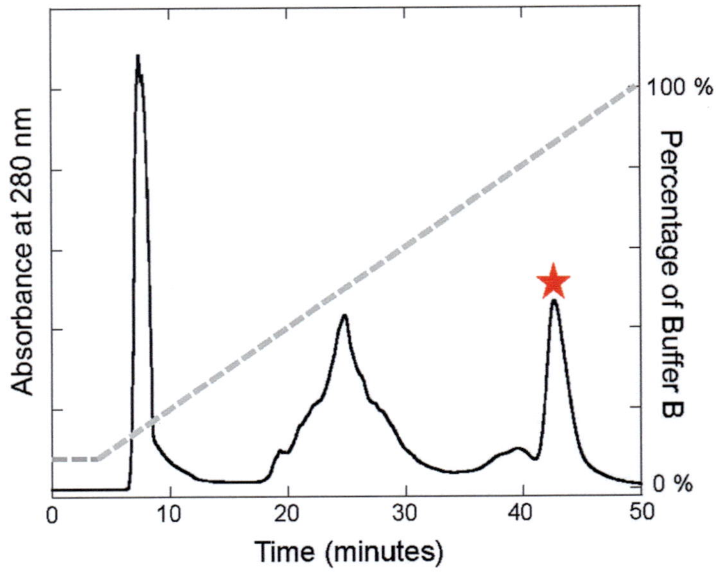

Fig. 2 Reversed-phase HPLC trace for the purification of $Cav1_{62-178}$. The desired peak is indicated with a red star

3.6 Caveolin-1 Reconstitution for NMR Studies

1. Approximately 6 mg of lyophilized protein was reconstituted into 600 μL of NMR buffer (*see* **Note 13**).
2. Samples were vortexed and briefly warmed in a 90 °C water bath until clear and filtered through a 0.2 μm spin filter.
3. The protein sample is now ready for NMR analysis. Figure 3 shows representative data for caveolin-1 (62–178).

4 Notes

1. The native methionine residue (M111) found within caveolin-1 (62–178) was mutated to a leucine residue to avoid cleavage by cyanogen bromide.
2. If full-length caveolin-1 is desired, both M32 and M111 need to be mutated to leucine residues.
3. The C-terminus of the TrpLE protein encodes for a thrombin cleavage site (LVPRGS) that may be used as an alternative to cyanogen bromide cleavage.
4. The three cysteines in caveolin-1 are natively palmitoylated. However, mutation of all three cysteines to serines did not affect its localization in the cell [13]. Therefore, it is likely that palmitoylation has a limited effect on the caveolin-1 fold.
5. Do not harvest cells at 4 °C!
6. Excessive foaming can occur if the stirring speed is too high.

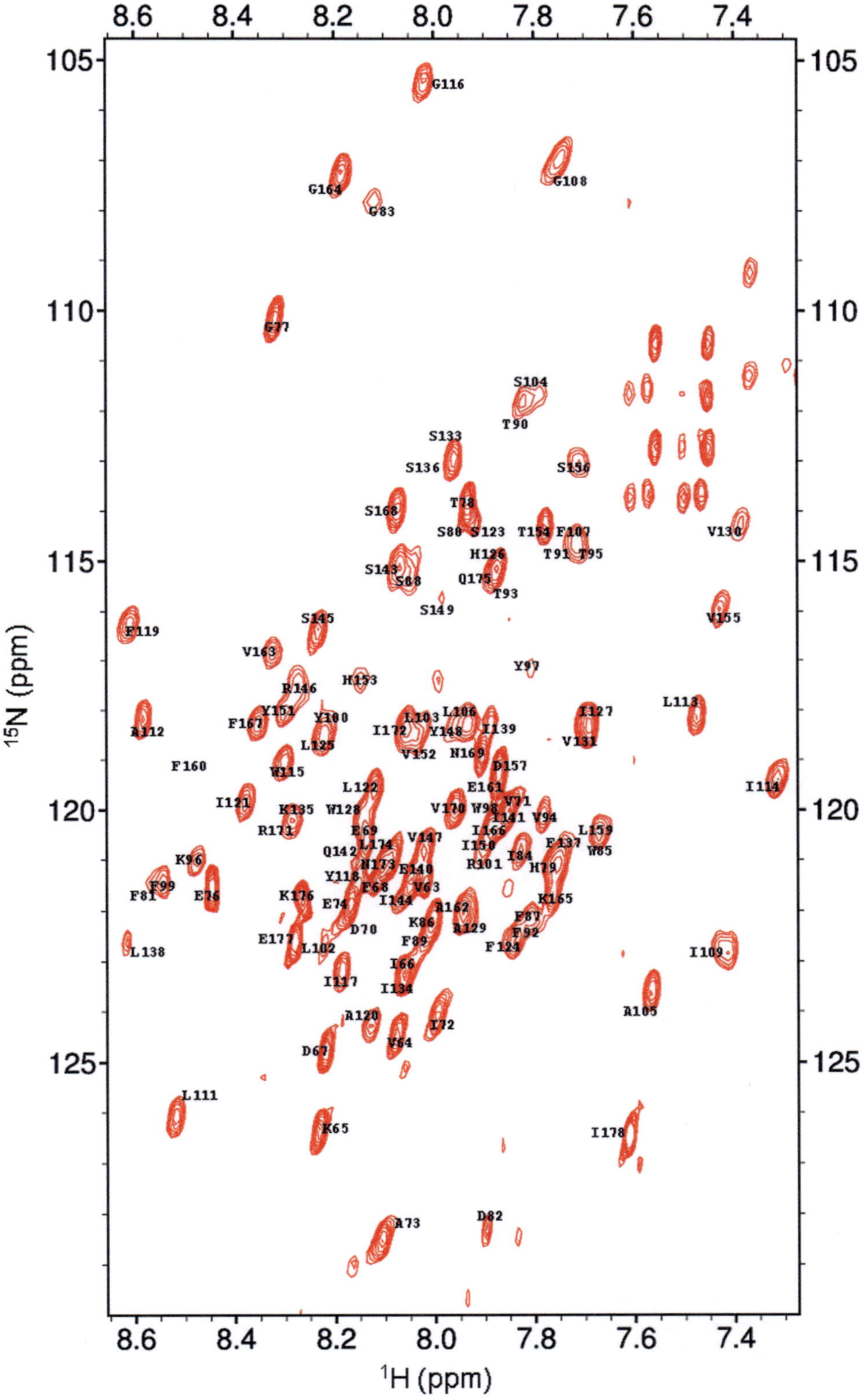

Fig. 3 Assigned ^{1}H,^{15}N TROSY-HSQC (transverse relaxation-optimized spectroscopy, heteronuclear single quantum coherence spectroscopy) spectrum of Cav1$_{62–178}$. The spectrum was acquired with 256 complex points in t1 (^{15}N) and 2048 complex points in *t*2 (^{1}H) [11]. NMR data were collected on 600 MHz Bruker Advance II spectrometer equipped with a cryoprobe

7. Given that inclusion bodies are gray in color, while unlysed *E. coli* cells are typically brown, if the pellet contains an appreciable amount of unlysed cells, an additional detergent buffer wash is necessary.
8. To remove all remaining supernatant, invert the tube on a clean paper towel for 5 min.
9. After each of the centrifugation steps, the pellet can be stored at −80 °C until needed.
10. After washing the cell pellet with the Sucrose, Detergent, and IPA buffer, the TrpLE-M-Cav1$_{62-178}$ band (~26 kDa) should be the predominant band observed on the gel. If this is not the case, more Detergent buffer washes are necessary.
11. After removal of the formic acid, a flaky, transparent film should be present.
12. Depending on the size of the dried pellet, it may be necessary to utilize larger volumes to dissolve the pellet. However, only 4.5 mL should be loaded onto the HPLC column at a time to achieve optimal protein recovery.
13. Initially, the solution will be slightly cloudy, but as it is warmed, it will become clear.

Acknowledgement

This work was supported by the National Institutes of Health (NIH-R01 GM093258 to K.J.G.).

References

1. Song Y, Mittendorf KF, Lu Z et al (2014) Impact of bilayer lipid composition on the structure and topology of the transmembrane amyloid precursor C99 protein. J Am Chem Soc 136:4093–4096
2. Ng HQ, Kim YM, Huang Q et al (2012) Purification and structural characterization of the voltage-sensor domain of the hERG potassium channel. Protein Expr Purif 86:98–104
3. Chen L, Lai C, Lai J et al (2011) Expression, purification, detergent screening and solution NMR backbone assignment of the human potassium channel accessory subunit MiRP1. Protein Expr Purif 76:205–210
4. Tian C, Vanoye CG, Kang C et al (2007) Preparation, functional characterization, and NMR studies of human KCNE1, a voltage-gated potassium channel accessory subunit associated with deafness and long QT syndrome. Biochemistry 46:11459–11472
5. Kim HJ, Howell SC, Van Horn WD et al (2009) Recent advances in the application of solution NMR spectroscopy to multi-span integral membrane proteins. Prog Nucl Magn Reson Spectrosc 55:335–360
6. Parton RG, Hanzal-Bayer M, Hancock JF (2006) Biogenesis of caveolae: a structural model for caveolin-induced domain formation. J Cell Sci 119:787–796
7. Root KT, Plucinsky SM, Glover KJ (2015) Chapter nine - recent progress in the topology, structure, and oligomerization of caveolin: a building block of caveolae. In: Kenworthy A (ed) Lipid domains. Current topics in membranes, vol 75. Elsevier, New York, pp 305–336
8. Le P, Guay G, Altschuler Y et al (2002) Caveolin-1 is a negative regulator of caveolae-mediated endocytosis to the endoplasmic reticulum. J Biol Chem 277:3371–3379

9. Ju H, Zou R, Venema VJ et al (1997) Direct interaction of endothelial nitric-oxide synthase and caveolin-1 inhibits synthase activity. J Biol Chem 272:18522–18525
10. Lee H, Woodman SE, Engelman JA et al (2001) Palmitoylation of caveolin-1 at a single site (Cys-156) controls its coupling to the c-Src tyrosine kinase - targeting of dually acylated molecules (GPI-linked, transmembrane, or cytoplasmic) to caveolae effectively uncouples c-Src and caveolin-1 (Tyr-14). J Biol Chem 276:35150–35158
11. Plucinsky SM, Glover KJ (2015) Secondary structure analysis of a functional construct of caveolin-1 reveals a long C-terminal helix. Biophys J 109:1686–1688
12. Studier FW (2005) Protein production by auto-induction in high density shaking cultures. Protein Expr Purif 41:207–234
13. Dietzen DJ, Hastings WR, Lublin DM (1995) Caveolin is palmitoylated on multiple cysteine residues. Palmitoylation is not necessary for localization of caveolin to caveolae. J Biol Chem 270:6838–6842
14. Marley J, Lu M, Bracken C (2001) A method for efficient isotopic labeling of recombinant proteins RID A-7811-2009. J Biomol NMR 20:71–75
15. Diefenderfer C, Lee J, Mlyanarski S et al (2009) Reliable expression and purification of highly insoluble transmembrane domains. Anal Biochem 384:274–278

Chapter 14

Tagging and Deleting of Endogenous Caveolar Components Using CRISPR/Cas9 Technology

Elena Shvets and Carolina Mendoza-Topaz

Abstract

Here, we describe how to utilize CRISPR/Cas9 technology in the generation of tissue culture cells with fluorescently tagged caveolar components as well as cells deleted of endogenous caveolar components. As one example, we will describe tagging of EHD2, caveolar neck protein, with Green Fluorescent protein (eGFP) from endogenous loci (knock-in, KI). As another example, we will describe deletion (knock-out, KO) of Caveolin1 (Cav1), an essential caveolar component in NIH/3T3 cells. In both instances, the modifications were achieved by using Cas9 delivery on plasmid DNA by electroporation and by utilizing FACS cell sorting for selection or enrichment of edited population of cells. We also provide a list with tested gRNA sequences to successfully produce KI and KO of other caveolar components.

Key words Caveolae, Caveolin1, Cavin1, EHD2, Genome-editing, Knock-in, Knock-out, NIH/3T3 cells

1 Introduction

Genome-editing with CRISPR/Cas9 has been broadly used in recent years and its applications, modifications, and alternative uses are an active field of research. The basic technology includes the introduction of a desired change to the genomic DNA by co-expression of Cas9 protein together with small gRNA molecule complementary to the target sequence of interest. Cas9 is an enzyme whose original function was to participate in immunological defense of certain bacteria against DNA viruses; however, recently it has been adopted for genome-editing application in mammalian cells and other cellular model systems.

To engineer mammalian genomes using CRISPR/Cas9, the components can be delivered to the cells either as a ribonuclear complex, a plasmid DNA, or mRNA molecules. The method described here uses plasmid DNA. Upon expression by host machinery, Cas9 protein is directed toward the specific genomic region, in a process mediated by the gRNA, thus leading to

Cedric M. Blouin (ed.), *Caveolae: Methods and Protocols*, Methods in Molecular Biology, vol. 2169,
https://doi.org/10.1007/978-1-0716-0732-9_14,

genomic cleavage by Cas9 in this particular region. If no homologous DNA is found, this genomic cleavage will be repaired by "non-homologous end joining" (NHEJ) resulting in stochastic insertions and deletions (indels) of nucleotides at the target locus. For insertion (knock-in, KI) of any extra sequence in the required endogenous locus, such as the fluorescent tag, it is necessary to include in the reaction another component that acts as a "donor DNA", so that the genomic DNA can be repaired by "homology-directed repair" (HDR). This "donor DNA" is actually a sequence homologous to the region of interest, thus containing homology arms both upstream and downstream to the cleavage site, induced by Cas9. If the donor DNA contains an "extra" sequence encoding a fluorescent tag, for example, it would be incorporated into the genome in this particular location. The efficiency of KI mediated by HDR is substantially lower than KO mediated by NHEJ, therefore there is a need of an appropriate selection method or enrichment for those rare events.

Caveolar assembly and kinetics are strongly affected by overexpression of its components [1, 2]. For example, overexpression of Cav1-GFP causes aberrant accumulation in endosomal compartments that may be mistaken for caveolae-positive endosomes [3, 4]. Hence, the possibility to follow caveolae in live cells using fluorescent tags inserted into endogenous locus, expressed under the endogenous promoter, offers an invaluable tool to properly study caveolae and their dynamics. A similar discrepancy in performance between overexpression and endogenous proteins might be general for other protein complexes and dynamic organelles.

Here, we describe the generation of endogenous EHD2 tagged with fluorescent tag EGFP at its C-terminus as an example of KI and generation of Cav1 KO as an example of gene deletion in NIH/3T3 cells. Additionally, we provide a list with gRNAs that resulted in a confirmed genomic modification to generate endogenous Cavin1-mCherry, EHD1-GFP, EHD4-GFP, Cavin2-GFP, GFP-Pacsin2, as well as Cav1 KO, EHD2 KO, EHD1 KO, and EHD4 KO cell lines (all done in mouse genome, NIH/3T3 cell line (2, 5, and unpublished results)) (*see* **Note 1**). The approach is based on plasmid DNA electroporation into mammalian cells, enrichment of the engineered cells by cell sorting, and finally phenotypic screening based on either the absence of the protein by Western Blot analysis (knock-out, KO) or colocalization of a fluorescent protein with Cav1 antibody (knock-in, KI). In the case of KO, the pool of cells transfected with the plasmids is sorted soon after gene delivery, for the enrichment of population that was efficiently transfected with Cas9-GFP plasmid. In the case of KI, the selection is done long after transfection by FACS sorting aiming for the fluorescent protein encoded in the donor DNA (to avoid selecting a population with plasmid DNA integration into the genome) (*see* **Note 1**). The selected clones are later validated by genomic PCR and sequencing (Table 1).

Table 1
Sequences used for the generation of KI modification

KI	gRNA	Sequences donor DNA left arm	Sequences donor DNA right arm
EHD2-GFP	GTTGGCTACACGGCTACACG	GGATAATGCAACTCGTGCTTAAAG TTCAGCAGAGCCCTTCTGTC	GACAGCAGCGACTGCTGG CAATGCCCCAGGATATCCATG
CAVIN1-mCherry	GCTGGAGATCACCGAGGAGTCGG	AACTTCTAAACTCTCCCAAACTG GTCGCTGTCGCTCTTGT	GCAGGACTCGCAGGGCTCTGCCC CATGGAAAACAAGGCTTCC
CAVIN2-GFP	GGGCCAGTGCATCCACCTCG	CCTTGCCAGTTAAGCTTCATGG GGCAGTCTGATCCACATGCA	GTGCACAGCCAGAACACTGG GAATGGAAAGGTCACTTCAAAGTG
GFP-PACSIN2	GTCACCTACGATGACTCTGT	AACCTAGTGTTTCTCATCTTTGGG TTCTGCAGAGGACGAGGGG	ATGTCTGTCACCTACGATGACTC CCTGATCACTAGGTTTCCTGTAA

Given sequences represent gRNA used in this study as well as sequences used for cloning of donor DNA. Please note that the actual primers for cloning of donor DNA contained additional sequence complementary to other fragments used in its assembly (approach required for cloning using Gibson assembly method)

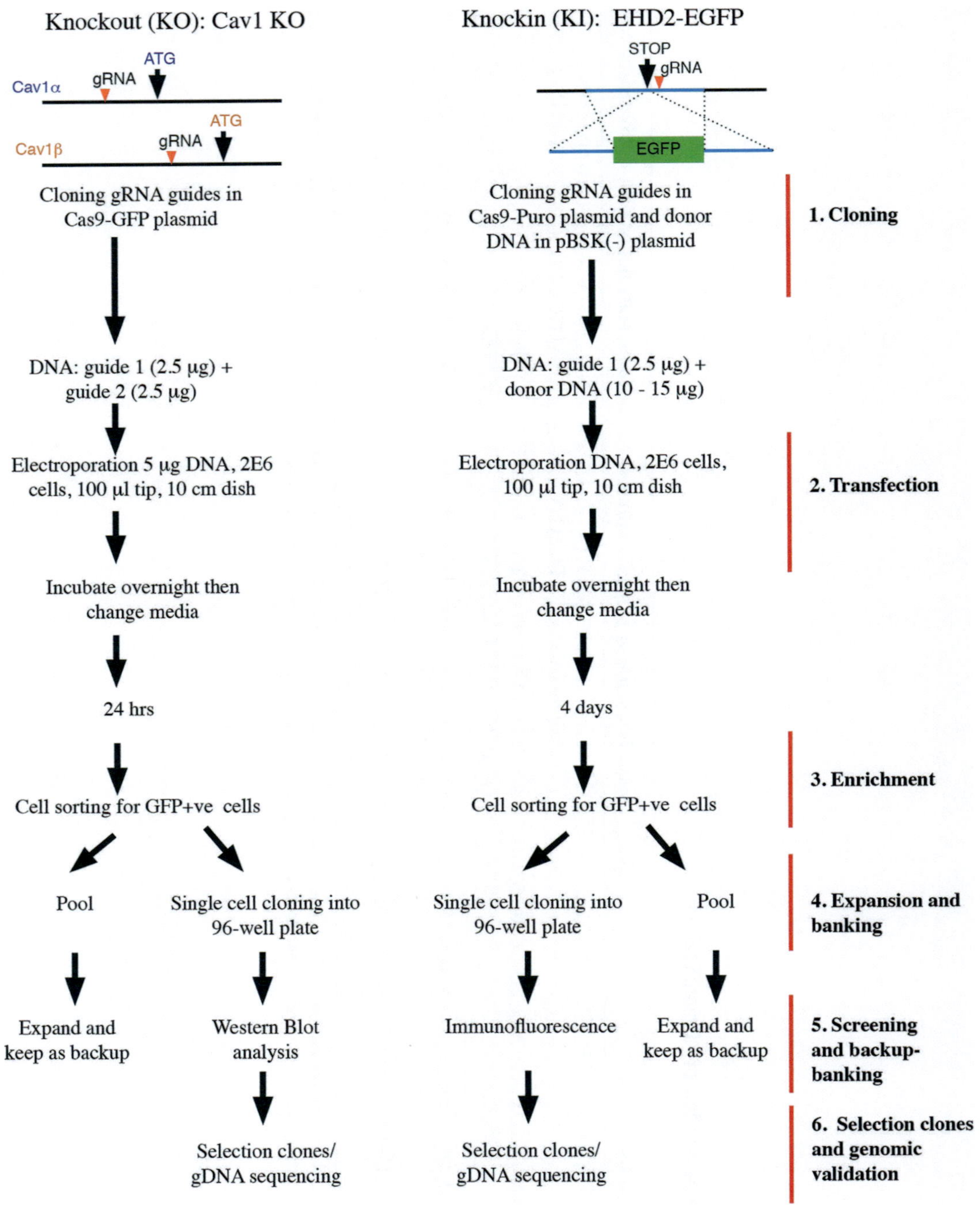

Fig. 1 Schematic presentation of the workflow described in this protocol to obtain either cells deleted of caveolar components (left panel) or cells expressing fluorescently tagged endogenous caveolar protein (right panel)

The protocols and methods described in this chapter consist of the following steps (Fig. 1):

1. Cloning of the guides (gRNA) into Cas9 plasmids and cloning of the donor DNA (*see* **Note 2**).

2. Gene delivery into mammalian cells by electroporation.
3. Enrichment of the edited cells by cell sorting.
4. Expansion of the cells (as single cell clones or as a pool) and banking of cells.
5. Screening of selected clones by Western Blot analysis for KO detection and by immunofluorescence for KI detection.
6. Expansion and banking of selected clones and genomic characterization.

2 Materials

2.1 Plasmids and Primers

1. Cas9 plasmids pSpCas9(BB)-2A-GFP (PX458) for generation of knockout cell-lines or pSpCas9(BB)-2A-Puro (PX459) for generation of KI cell-lines (Addgene plasmid #48138 and #48139) [6] (*see* **Note 2**).
2. Primers and pBlueScript SK (−) plasmid for generation of donor DNA (*see* **Notes 3–5**). Inserts: 1 kb of genomic DNA sequence on either side of mouse EHD2 stop codon and DNA coding for EGFP (*see* Table 1 for a list of sequences used for generation of donor DNA in this study).
3. gRNA, *see* Tables 1 and 2 (*see* **Notes 6** and **7**).

2.2 General DNA Manipulation

1. Agarose gel, DNA molecular weight marker, SYBR safe, electrophoresis chambers, and power pack.
2. Gibson Master kit assembly (New England BioLabs) or other assembly reagents such as GeneArt, In-Fusion NEBuilder, 37 and 50 °C heating block, gel extraction kit, miniprep and midiprep DNA purification kits, 1.5 mL tubes, scalpel, and UV transilluminator.

Table 2
Sequences used for the generation of KO modification

KO	gRNA1a	gRNA2
EHD1	CAAGGATGCCCGCCGCAAGA	CATCCGCCACCTGATCGAGC
EHD2	TGCGGATGGCCTCAGGTCGC	CGAAAGTCACCATGTTCAGC
EHD4	CTGGCGGGCGCGAGCGCTC	CTGGTGGGCCAGTACAGCA
CAV1	CAGACATGCTGGCGCGTGGC	AGTGTATGACGCGCACACCAAGG
PACSIN2	TACCTCCCAGAAGCTGTCGC	GTCACCTACGATGACTCTGT

Given sequences represent a pair of gRNA used in this study for the deletion of the target gene

3. 100% and 70% EtOH, 3M $NaCH_3CO_2$ pH 5.2. Mix and adjust pH with HCl. Chemically competent bacteria for transformation with superior efficiency.

2.3 Tissue Culture

1. Low-passage NIH/3T3 cells (<15) (*see* **Note 8**).
2. DMEM medium supplemented with penicillin/streptomycin and 10% of fetal bovine serum (FBS) (*see* **Note 9**).
3. 37 °C, 5% CO_2 incubator, and 1× trypsin-EDTA solution.
4. Plasticware: 10 cm tissue culture dishes for culturing, 96-well tissue culture plates flat bottom for sorting and growing individual clones, 24- and 6-well plates for screening and expansion of selected clones, and 50 and 15 mL tubes.
5. Multichannel pipette, sterile disposable reservoir for medium.

2.4 Electroporation

1. Neon electroporation system; 100 μL tips, medium without antibiotics.
2. 50 mL falcon tubes, 1.5 mL Eppendorf tubes.
3. Centrifuge, vortex.

2.5 Genomic DNA Purification and PCR Reaction

1. Cell lysis and protein precipitation solution (QIAGEN) or other commercially available Genomic DNA purification solution.
2. Reagents for PCR reaction: buffer, dNTPs, enzyme KOD Hot Start DNA Polymerase (or any other proofreading Polymerase enzyme), primers, DNA template, and thermocycler machine.
3. Agarose gels, DNA ladder, and 1× TAE buffer.

2.6 FAC Sorting

1. PBS and trypsin-EDTA solutions.
2. Sorting tubes.
3. 50 μm cell strainers.
4. Sony iCyt Synergy Dual Channel High-Speed Cell sorter/ Beckman Coulter MoFlo High-Speed Cell sorter (or any other compatible cell sorter).

2.7 Cell Lysis

1. PBS, 2 mL tubes, and ice.
2. Rubber or silicone cell scrapers with pivoting blades.
3. 40% *n*-octylglucoside (OG) stock (stock aliquoted and kept at −20 °C).
4. Triton X-100.
5. EDTA-free protease inhibitors cocktail.
6. Lysis buffer: 50 mM Tris pH 8, 300 mM NaCl, 5 mM EDTA, 1% Triton X-100, 1% *n*-octylglucoside (OG), and complete

protease inhibitor cocktail. Prepare 50 mL without OG and keep it at 4 °C. Add 250 μL of 40% OG per 10 mL lysis buffer just before use.

2.8 Protein Analysis by Western Blot

1. 4–12% Bis-Tris SDS-PAGE gels, 1 mm thick, 15 wells.
2. Power pack.
3. Pre-stained protein molecular weight marker.
4. 1× MES SDS running buffer. Prepare 20× stock as follows: 97.6 g MES, 60.6 g Tris base, 10 g SDS, and 3 g EDTA dissolved in 400 mL of MPW, mix well and make final volume to 500 mL. Store at 4 °C covered in foil to protect from light.
5. Whatman filter paper, PVDF membranes, methanol, and transfer tank for wet transfer.
6. Transfer buffer. Prepare 20× stock (without methanol) as follows: 30 g Tris base and 144 g glycine in 1 L. For 2 L, 1× transfer buffer, mix 200 mL 20× Tris/glycine stock, 400 mL methanol, and make final volume to 2 L.
7. PBST solution: 0.1% Tween-20 in 1× PBS.
8. Blocking solution: 5% dried skimmed milk powder in PBST.
9. Shaker for blots.
10. DTT and 4× SDS-sample buffer.
11. Antibodies: rabbit anti-Cav1 (BD Biosciences 610060), rabbit anti-Cavin1 (Abcam ab48824), goat anti-EHD2 (Abcam ab23935), any anti-RFP and anti-GFP antibodies suitable for immunofluorescence and Western Blot and any anti-tubulin (or other housekeeping gene), anti-mouse HRP, and anti-rabbit HRP for Western Blot analysis.
12. iBlock (or other protein-based blocking reagent).
13. Immobilon Western Chemiluminescent HRP Substrate (Millipore) or ECL Western Blot Detection Reagent Kit GE Healthcare.
14. Cassettes for X-ray films.

2.9 Immunofluorescence

1. Sterile 2 mm × 22 mm coverslips to use inside each well of a 24-well plate and glass slides to mount them or 8-well chambers (like ibidi chambers).
2. Fixation solution: 4% PFA in PBS.
3. Blocking and permeabilization solution: 2% saponin, 10% FBS in PBS.
4. Antibodies: rabbit anti-Cav1 (use 1:3000), goat anti-EHD2 (use 1:2000), mouse anti-GFP (use 1:2000), and fluorescent conjugated secondary antibodies (like anti-goat Alexa 555, anti-rabbit Alexa 488, and anti-mouse Alexa 647) (use all 1:500).

3 Methods

Carry out all the DNA and protein procedures in a laboratory bench at room temperature unless otherwise specified and tissue culture work in the tissue culture hood with sceptic technique. Cells are incubated in a 37 °C incubator with 5% CO_2. All centrifugations are carried out at room temperature unless otherwise specified.

3.1 Donor DNA Constructs for EHD2-EGFP KI

1. Purify genomic DNA serving as a template for the generation of donor DNA constructs from NIH/3T3 cells using commercially available genomic DNA purification kits. Optionally, commercially available genomic DNA of relevant origin can be used for template generation (*see* **Note 10**).
2. Clone donor DNA construct by traditional molecular biology methods. Amplify approximately 1 kb of genomic DNA sequence on either side of the EHD2 stop codon from genomic DNA at sequences listed in Table 1. Amplify DNA coding for EGFP originated from pEGFP-N1 (or any of its variants). The stop codon of the EHD2 to be deleted and the coding fragment of the fluorescent protein to be fused in frame with linker as described [5] and inserted into pBlueScript SK(−) using Gibson Master kit assembly (New England BioLabs) according to the manufacturer's instructions (*see* **Note 5**).

3.2 Cloning gRNA into Cas9 Vector

Clone gRNA with a simplified version of the original protocol described in Ran et al. [6]. Briefly, simply combine BbsI digested vector and ligation in the same step (based on the "ELAN" method described by Cost and Cozzarelli [7]). This eliminates oligo kinase, vector phosphatase, and gel purification steps described in the original Nature Protocol.

Start with annealed gRNA oligos with overhangs as described by Zhang Lab (do not phosphorylate oligo).

1. Anneal top and bottom gRNA oligos:
   1 μL oligo 1 (100 μM).

   1 μL oligo 2 (100 μM).

   8 μL ddH_2O.

   95 °C for 5 min and keep on the bench to cool down and anneal.
2. Digest pX458 and pX459 vectors with BbsI (do not dephosphorylate treat):

   2 μL 10× NEB Buffer 2.1.

   16 μL H_2O.

   1 μL plasmid (1 μg or less).

1 μL BbsI.

37 °C for 1 h.

3. Add ligation components directly to the digestion product:
   2.5 μL 10× T4 Ligase Buffer.

   1 μL annealed oligo (10 μM stock; 0.4 μM final concentration).

   1.5 μL T4 DNA ligase.

   37 °C for 1 h.

4. Transform competent bacteria with 2 μL directly from the above reaction.
5. Pick 3–4 colonies, do mini-prep, and verify by sequencing. Primer for sequencing: U6-Fwd: GAGGGCCTATTTCCCATGATTCC. The efficiency is usually very high (90–100%) so there is no need to screen by colony PCR or pick more than 3–4 colonies per cloning.

   Using the method described above, clone gRNAs for Cav1 KO into pSpCas9(BB)-2A-GFP (pX458) and clone gRNAs for EHD2-GFP KI into pSpCas9(BB)-2A-Puro (pX459). Those guides were originally designed using Feng Zhang lab software at http://crispr.mit.edu/ (for sequences, *see* Tables 1 and 2) (*see* **Note 11**).

3.3 Midiprep and Concentration of DNA Constructs

1. Purify plasmids DNA using midi-prep kit according to the manufacturer's instructions.
2. Concentrate DNA (using NaAc/ethanol or other widely used method) if obtained yield below 500 ng/mL, resuspend DNA pellet in 10 mM Tris-Cl, pH 8.5 to bring to final DNA concentration to 1–1.5 μg/mL (*see* **Note 12**).

3.4 Electroporation

1. Day 1: Thaw a low-passage cryotube of NIH/3T3 (*see* **Note 13**) at 37 °C and transfer the cells to a 10 cm tissue culture dish with 10 mL of pre-warmed (37 °C) DMEM media containing 10% non-heat-inactivated FBS and pen/strep. Leave cell to recover overnight at 37 °C in the incubator.
2. Day 2: the cells should be approx. 90% confluent. Prepare pre-warm culture media, Trypsin/EDTA, and PBS (without Ca^{2+}/Mg^{2+}). Aspirate media from cells, wash with PBS and add 1 mL of warm Trypsin/EDTA, and incubate at humidified 37 °C/5% CO_2 incubator for 2–3 min. Resuspend trypsinized cells with 9 mL of warm culture media, transfer 3 mL into new dish containing 7 mL of fresh culture media. Mix well and incubate cells overnight for further proliferation.
3. Day 3 or 4: passaged cells should be approx. 80–90% confluent. Plan one 10 cm plate/one sample of electroporation ($2–3 \times 10^6$ cells/sample).

4. Prepare pre-warm normal culture media and culture media without antibiotics to seed electroporated cells. Trypsin/EDTA and PBS (without Ca^{2+}/Mg^{2+}). Aspirate media from cells, wash with PBS and add 1 mL of warm Trypsin/EDTA, and incubate at humidified 37 °C/5% CO_2 incubator for 2–3 min.
5. Remove cells from incubator and harvest them in 7–10 mL of normal serum-containing media, pipette cells up and down to mix.
6. Remove small proportion of trypsinized cells and count them.
7. Centrifuge cells at 300 × *g*, 5 min at room temperature.
8. In the meantime, prepare 10 cm dishes with 10 mL of culture, serum-containing medium without antibiotics, and incubate plates in a humidified 37 °C/5% CO_2 incubator.
9. Aspirate media and resuspend cells with pre-warm PBS.
10. Centrifuge cells at 300 × *g*, 5 min at room temperature.
11. In the meantime, prepare DNA in 1.5 mL tubes as follows:
    (a) For generation of KO cells: relevant gRNA1 cloned into pX458 together with gRNA2 cloned into pX458, use 2.5–5 μg of each DNA construct (*see* **Note 14**).
    (b) For generation of knock-in cells: prepare relevant gRNA cloned in pX458 (2.5–5 μg/sample) together with donor DNA cloned into pBlueScript SK (−)- vector (10–15 μg/sample).
    (c) Aspirate PBS and resuspend cells in 120 μL/sample of Resuspension Buffer R at a final density of 1.0×10^7 cells/mL (*see* **Notes 15** and **16**).
12. Set up Neon electroporation system according to manufacturer's instructions.
13. Use the next settings for electroporation with 100 μL tips: pulse voltage 1400 V, pulse width 20 ms, and pulse number 2 (protocol that works well for NIH/3T3) (*see* **Notes 16–18**).
14. Electroporate cells according to the manufacturer's instructions.
15. Immediately after electroporation, release cells into a plate with pre-warmed media without antibiotics and place into a humidified 37 °C/5% CO_2 incubator.
16. Optionally, release 100 μL of electroporated cells into a tube with 1 mL of pre-warmed medium. Close lid and mix well by inversion and add it into one 10 cm dish, mix well, and incubate plates in a humidified 37 °C/5% CO_2 incubator.
17. Continue until all samples are processed.

18. Next day, change medium with pre-warmed medium with antibiotics (*see* **Note 19**).
19. FACS sorting of knock-out cells (*see* **Note 20**): 48 h following electroporation, trypsinize cells and wash once with PBS as described earlier, resuspend cells in PBS/2%BSA (or any other buffer recommended for FACS sorting) at a final density of 5–10 × 10^6 cells/ml (roughly 0.5 mL/10 cm plate sample), and filter with 50 μm cell strainer into sorting tubes.
20. Prepare 3× 96-well plates with 100 μL/well of culture media and keep warm at 37 °C until sorting.
21. Prepare a similar amount of untransfected cells to serve as a negative control for FACS setup and proceed with FACS sorting. Sort based on GFP expression one viable cell/well into 3× 96-well plates for each KO sample (*see* **Note 21**).
22. Collect the pool of remaining transfected cells into a collection tube with 1 mL of culture media and seed them into 10 cm dish. This pooled sample will serve as a backup option in case of failure to recover clones from single cells, as it can be used later to isolate single clones using limiting dilution or new round of FACS sorting (*see* **Note 22**).
23. FACS sorting of knock-in cells (*see* **Note 19**): 5 days following electroporation process, cells as described before. Prepare 12-well or 6-well plates with warm culture media for cells collection. Sort pooled population of GFP expressing cells (or any other fluorescent protein used). If cells exhibit a clear pattern of different expression levels, consider sorting different subpopulation into different pools (*see* **Note 22**).
24. Add sorted cells to wells of 12-well plate or 6-well plate, according to the number of positive cells being sorted and incubate at a humidified 37 °C/5% CO_2 incubator.

3.5 Growing, Expansion, Selection, and Banking of Clones

Single clones

1. 3–4 days following sorting of single cells, add 100 μL warm medium to each well in the 96-well plate.
2. Check 96-well plates to confirm positive outgrowth of single clones.
3. Approximately 10 days after sorting, select wells in 96-well plates where cells grew into large colonies. Exclude wells where two colonies are clearly observed.
4. Transfer clones into 24-well plates for further proliferation (*see* **Notes 23** and **24**).
5. When cells in most wells are 50–70% confluence, split colonies into 12-well plate for analysis by WB and/or 24-well plate

containing coverslips for IF (or 8- or 16-well ibidi chambers for imaging).

6. Transfer selected clones into 6-well plate for further proliferation and analysis. Cells can be splitted when confluent enough into replica 6-well plate for freezing.
7. Freeze selected clones several times. Consider freezing a certain number of potential clones before the end of the analysis (as the analysis might take long time).
8. Verify the viability of frozen cells, by testing recovery after thawing of selective samples.

Pooled cells

1. Closely follow cells recovery on a daily basis, change media every 2–3 days, split cells when reach 70–80% confluence into 12-well plates for analysis by WB and/or IF.
2. Freeze down recovered cells in few different batches.
3. Verify the viability of frozen cells, by testing recovery after thawing of selective samples.

3.6 Screening

Western blot analysis and immunofluorescence are two primary techniques to screen for positive clones and also validate proper tagging and localization of tagged proteins. It will also provide valuable information on whether all alleles are deleted or tagged and whether the tagged protein is of the expected size.

Western blot

1. Harvest 12-well plate with confluent clones and lyse cells with 50 μL of complete lysis buffer.
2. Run 20 μL of each sample in 4–12% Bis-Tris SDS-PAGE gels, including un-transfected samples and pre-stained protein marker. In the case of knock-in cell-line, include the sample with transiently expressed GFP tagged protein for positive control. Similarly, if the specificity of the detection antibody is not clear, include samples with different homologues of tested protein. For example, include overexpressed EHD1, EHD2, and EHD4 proteins when testing the editing of EHD2 protein.
3. Transfer separated proteins from the gel into PVDF membrane, block for unspecific signal, and incubate membranes with relevant primary antibody: rabbit anti-Cav1 antibody 1:10.000 (for Cav1 KO), goat anti-EHD2 (for EHD2 KO and KI) in 5% milk PBST for 2 h at room temperature or overnight at 4 °C, wash 5 × 5 min with PBST and then incubate 1 h at room temperature with relevant HRP-conjugated secondary antibody.
4. Wash 5× 5 min with PBST and develop with a chemiluminescent reagent.

5. Re-probe PVDF membrane with anti-GFP or and anti-tubulin antibody. Consider running several gels in parallel if probing with more antibodies is needed for full analysis (for example, probing EHD2 KO clones with anti-EHD1 and anti-EHD4 antibodies).

Immunofluorescence

1. Perform staining in your favorite format, the more samples you can easily handle the better. Eight or 16-well chambers, like ibidi, are recommended as they can be imaged directly without the need of mounting. Another good option is to seed the cells directly on coverslips in 24-well plate.
2. Wash cells thoroughly three times with PBS.
3. Fix cells with 4% PFA for 15 min at room temperature.
4. Wash with PBS.
5. Block and permeabilize cells with 0.2% saponin, 10% FBS in PBS for 30 min at room temperature.
6. Add primary antibody mix, for example, rabbit anti-Cav1 (1:3000), goat anti-EHD2 (1:2000), or mouse anti-GFP (1:2000) in PBS/10% FCS blocking solution. Incubate 2–4 h at room temperature.
7. Wash 4× with PBS/2% FCS.
8. Add secondary antibody mix accordingly and incubate 1 h at room temperature in humid chamber (*see* **Note 25**).

3.7 Validation of Selected Clones

1. PCR the edited genomic DNA region.
2. Purify the PCR product using any commercially available kits.
3. Perform Sanger sequencing of the PCR product (*see* **Note 26**).

4 Notes

1. When tagging endogenous protein, it is utmost important to assure that this modification does not affect protein function/structure. In case of Cav1, Cavin1, EHD1, EHD2, and EHD4 insertion of fluorescent protein did not affect their solubility, ability to bind their interaction partners in same stoichiometry, as endogenous proteins, as well as localization of tagged proteins to caveolae. Cavin2 and Pacsin2 KI efficiently result in colocalization with Cav1 and Cavin1 but they were not analyzed thoroughly (unpublished results, Fig. 2). Pacsin2 was tagged with eGFP at its N-terminus, while Cavin2 was tagged at its C-terminus—the choice of termini for every tagging was made based on known literature describing cloning and expression of exogenous protein from the plasmid. Partial tagging,

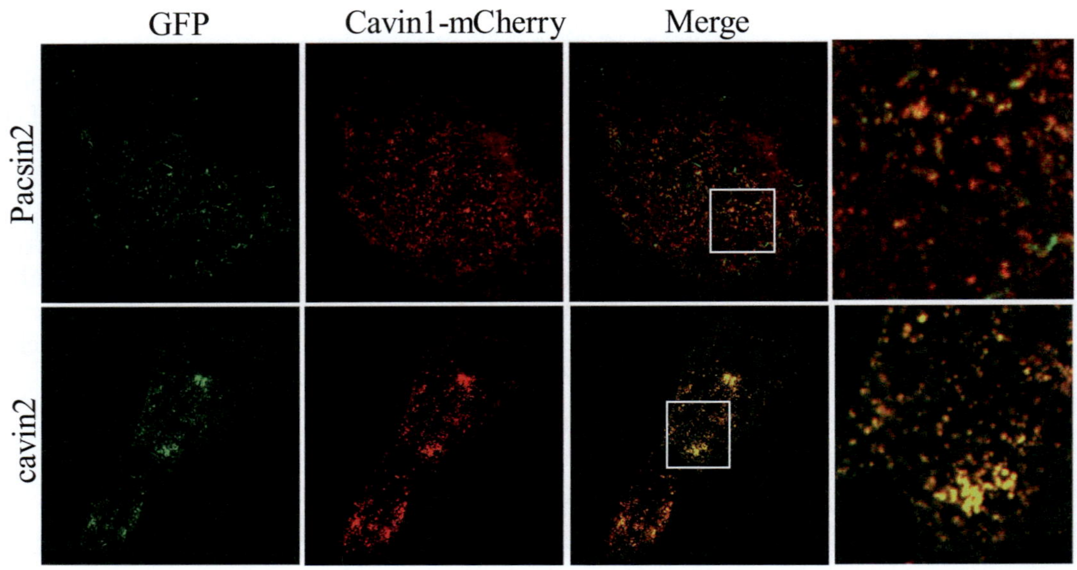

Fig. 2 Subcellular localization of two representative clones double KI of GFP-Pacsin2/Cavin1-mCherry (*see* **Note 27**) and Cavin2-GFP/cavin1-mCherry. All proteins were tagged with fluorescent proteins from endogenous loci. Selected clones were seeded in a 16-well iBidi chamber and imaged live using a confocal microscope. Please note substantial colocalization of cavin2 with cavin1 protein (as expected from literature) and partial colocalization between GFP-Pacsin2 and cavin1, while certain proportion of Pacsin2 protein shows non-caveolar, microspike-like localization, in agreement with previously published literature (de Kreuk et al. [10]). Bar = 10 μm

when only part of alleles are modified, can be beneficial, thus enables the visualization, yet minimizes chances of certain defects in protein function. In Mendoza-Topaz et al. [8], Cav1 KO was generated with one single gRNA (different from gRNA described here) but here we utilized two gRNA as this method proved to be more efficient and labor-saving when screening for KO clones.

2. Plasmid pX458 has sequence encoding for GFP that is co-expressed with Cas9 (via TA linker), thus indicating transfection efficiency and Cas9 expression level in target cells. pX459 has puromycin resistance that can be used for selection; however, as the target modification included insertion of fluorescent protein, this readout was used for selection, rather than puromycin resistance. Note that updated versions of these plasmids are now available.
3. It is crucial that donor DNA will be cloned into the backbone that has no mammalian promoter; otherwise, transfection will result also in transient expression of donor DNA that will mask on low-efficient tagging of endogenous protein.
4. In this study, Cas9 was delivered on plasmid, thus successful genome-editing required its expression in a given cellular

system. This strategy might prove to be problematic in a cellular system with low translation/transcription activity. Moreover, the disadvantages of this strategy will also include relatively long time until genome-editing can proceed, but also the longer presence of expressed Cas9 in transfected cells. In addition, there is a certain risk that Cas9 plasmid will be randomly integrated into the genome. Therefore, more recent commercially available systems that involve transfection with either Cas9 mRNA or with recombinant protein, assembled into RNP (RNA/protein) complexes, provide a significant advantage in controlling the timing of genome-editing, thus increasing efficiency and at the same time decreasing the risk for off-target modification.

5. The donor DNA, essential for the genomic insertion was designed with flanking regions of 1 kb size, to enable homologous recombination. However, more recent studies have shown that shorter arms can be efficiently used for CRISPR/Cas9 application [9]. Alternatively, synthetic donor DNA fragments (commercially available) can be considered to simplify this process.
6. Two sequences of gRNA for each area of cleavage were designed and tested. For EHD2 GFP tagging (at its C-terminal), gRNA was designed to be in closest proximity to stop codon (intended insertion site).
7. For KO generation, the modification relies on the introduction of possible mutation due to Cas9 cleavage and NHEJ repair. It is possible that one cleavage will be sufficient to result in such an effect; however, to optimize the outcome of protein deletion, rather than unnecessary mutation, certain rules should be followed: (1) The target for cleavage site downstream to first exons (exons 2–3) to minimize chances that alternative ATG will be used for transcription, yet do not target last exons, as it will result in truncation of protein rather than its deletion. (2) Target area that is common to all possible splice variants—this is relevant in case of Cav1 that exhibits alfa and beta isoforms (beta lacks 31 amino acids at it N-terminus). (3) Targeting the first 50% of the gene will also help to send an aberrant transcript with a premature stop codon (resulted from NHEJ-mediated INDELs) to Nonsense-mediated mRNA decay (NMD)-degradation of this transcript. (4) Target two different areas in the gene that will result in the deletion of gene fragment rather than mutation to increase chances for potential frameshift and consequential NMD. A different approach that involves HDR (homologous recombination) strategy can be also used for "hard to find" knockout targets. This approach involves designing donor DNA that will be used to replace part of the gene with fragment encoding for

fluorescent protein (or other marker/selection). Stop codon downstream to this insertion can be also introduced. The efficiency of this modification is lower than only cleavage-based strategy; however, screening for positive clones will be much more efficient as it will rely on FACS sorting of fluorescently positive cells. This strategy was also used for generation of EHD2 KO cells (unpublished data) with 90–100% efficiency score.

8. For the successful design of genome-editing experiments, take into careful consideration the characteristics of the chosen model system: expression level and function of target protein—if too low, protein tagging will not produce visible outcome; ploidy—most immortalized cell-lines exhibit polyploidy, thus making knock-out or mutagenesis modifications more difficult task. Cellular systems that contain single set of chromosomes can be considered a suitable model system.
9. Use non-heat-inactivated FBS for NIH/3T3 cells as significant differences in cell recovery and viability (especially following electroporation and in the outgrowth of single clones) were identified when using heat-inactivated FBS.
10. Since a certain degree of polymorphism can be observed within the same cell-line, especially as many constantly dividing cells are genetically unstable, it might be beneficial to use the template from in-house grown target cell-line for donor DNA.
11. The described tool from Zhang lab does not exist anymore, as multiple, more modern tools to design guides are readily available (some online tools are listed on http://crispr.mit.edu/). Design of good guiding sequences is of utmost importance, as it can save time and resources in target cell generation.
12. Electroporation technique can produce a very efficient gene delivery, but it requires DNA prep of very high quality and purity. Importantly, as the volume of DNA sample cannot exceed more than 10% of total sample volume (100 μL in this case), it is crucial to have highly concentrated DNA prep.
13. Use low-passage, highly viable cells for generation of genome-edited cell-lines.
14. If several proteins to be deleted at same cell-line (such as EHD1/EHD2/EHD4 triple KO), all three deletion setups (6 gRNA) can be mixed in the same reaction. This approach is faster and more importantly assures that randomly formed off-target defects are not "carried" from original clones, which could be the case when deletion is done at sequential, rather than simultaneous manner.
15. Proceed with electroporation as soon as possible as delaying it for more than 20–30 min will negatively affect cell viability.

16. For successful electroporation, avoid introducing air-bubbles into Neon tips, thus always allow 20% higher volume.
17. Note there are two types of tips in Neon device-10 μL and -100 μL. As gene tagging that relies on HDR that is not a very efficient process, it required working with a higher number of cells more suitable for 100 μL tips.
18. Note that as these settings have been optimized for NIH/3T3, you might find a different setup to be more suitable for your cells.
19. Optional: check fluorescence under a fluorescent microscope. Fluorescence will be detected in case of PX458 plasmid (relevant mostly in case of deletion with plasmid Cas9/GFP), while expression of endogenous tagged protein will take much longer (3–5 days) and might not be bright enough at this point.
20. For the generation of knockout cell, monoclonal population originated from single cell are isolated, since protein deletion requires editing of all alleles and no clear marker is available to detect such clones. Thus, sort cells based on transfection marker (co-expression of GFP with Cas9), and clones originated from single cells are later screened for deletion of Cav1 protein using WB and later confirming by antibody staining. Enrichment for transfected cells increases the chances of finding knockout cells within this population but does not guarantee it. For the generation of knock-in cells, the modification is done using a plasmid that does not have a fluorescent signal as transfection readout. Instead, the fluorescent signal reflects the desired genomic modification-insertion of fluorescent protein into genomic loci of endogenous protein, therefore sorting for such a signal is more specific. Random integration of the donor DNA with the fluorophore would not produce the fluorescent signal unless it is inserted in the correct frame under a promoter. Modification of only part of alleles is tolerated in case of endogenous tagging (as opposed to gene deletion) and can even be beneficial for the proper functioning of the modified protein.
21. Note that GFP expression with Cas9 is only transient; it will disappear after a few days. Do not sort for GFP if you detect it at a later stage, as these cells might stably express Cas9 due to random genomic integration.
22. Note that if unusually strong GFP signal is detected, it might be a sign of unspecific localization of this protein or expression of the mutated unfunctional tagged protein (due to random integration into off-target loci). Different levels of fluorescent signal might also reflect tagging of a higher number of alleles.

23. Consider the time that takes to trypsinize clones at 96-well plate, as cells are very fragile at this step, consider processing only a few colonies at the same time.
24. Choice of the number of clones to pick depends on the efficacy of genome-editing approach. Usually, the generation of KO is efficient, as it requires only Cas9-mediated cleavage, yet this process might not end up in protein deletion. Aim to have eventually at least 3–4 final positive clones/sample for your study.
25. For quick screening, one/two antibody might be sufficient.
26. It is possible that Cas9 cleavage will introduce only minor change/s in the gene sequence, not detectable by the apparent size difference, but the mutation may still cause to premature stop-codon incorporation and subsequent mRNA degradation.

References

1. Doyon JB, Zeitler B, Cheng J, Cheng AT, Cherone JM, Santiago Y, Lee AH, Vo TD, Doyon Y, Miller JC, Paschon DE, Zhang L, Rebar EJ, Gregory PD, Urnov FD, Drubin DG (2011) Rapid and efficient clathrin-mediated endocytosis revealed in genome-edited mammalian cells. Nat Cell Biol 13:331–337
2. Shvets E, Bitsikas V, Howard G, Hansen CG, Nichols BJ (2015) Dynamic caveolae exclude bulk membrane proteins and are required for sorting of excess glycosphingolipids. Nat Commun 6:6867
3. Hayer A, Stoeber M, Ritz D, Engel S, Meyer HH, Helenius A (2010) Caveolin-1 is ubiquitinated and targeted to intralumenal vesicles in endolysosomes for degradation. J Cell Biol 191:615–629
4. Parton RG, Howes MT (2010) Revisiting caveolin trafficking: the end of the caveosome. J Cell Biol 191:439–441
5. Yeow I, Howard G, Chadwick J, Mendoza-Topaz C, Hansen CG, Nichols BJ, Shvets E (2017) EHD proteins cooperate to generate caveolar clusters and to maintain caveolae during repeated mechanical stress. Curr Biol 27:2951–2962
6. Ran FA, Hsu PD, Wright J, Agarwala V, Scott DA, Zhang F (2013) Genome engineering using the CRISPR-Cas9 system. Nat Protoc 8:2281–2308
7. Cost GJ, Cozzarelli (2007) Directed assembly of DNA molecules via simultaneous ligation and digestion. Biotechniques 42:84, 86–89
8. Mendoza-Topaz C, Nelson G, Howard G, Hafner S, Rademacher P, Frick M, Nichols BJ (2018) Cells respond to deletion of CAV1 by increasing synthesis of extracellular matrix. PLoS One 13:e0205306
9. Zhang JP, Li XL, Li GH, Chen W, Arakaki C, Botimer GD, Baylink D, Zhang L, Wen W, Fu YW, Xu J, Chun N, Yuan W, Cheng T, Zhang XB (2017) Efficient precise knockin with a double cut HDR donor after CRISPR/Cas9-mediated double-stranded DNA cleavage. Genome Biol 18:35
10. de Kreuk BJ, Nethe M, Fernandez-Borja M, Anthony EC, Hensbergen PJ, Deelder AM, Plomann M, Hordijk PL (2011) The F-BAR domain protein PACSIN2 associates with Rac1 and regulates cell spreading and migration. J Cell Sci 124:2375–2388

Chapter 15

Pulling of Tethers from the Cell Plasma Membrane Using Optical Tweezers

Darius V. Köster

Abstract

Here, we describe how to extract tethers or lipid membrane nanotubes from the plasma membrane of cells using optical tweezers. This technique allows measuring the force required to hold the membrane tether at a constant length, which is related to the cell membrane tension. Following the evolution of this force during mechanical or chemical perturbations of the cell gives insight about the regulation of cell membrane tension. By pulling very long membrane tethers, one can also probe the membrane reservoir of a cell and a sudden rise in the tether force is usually due to the depletion of excess membranes stored in membrane folds or invaginations.

Key words Optical tweezers, Membrane tension, Membrane tether, Lipid nanotubes, Membrane mechanics

1 Introduction

The cell plasma membrane is intertwined and in constant interaction with the underlying cell cortex, which is continuously remodeling. This active, multicomponent system makes it difficult to directly assess the mechanical properties of the plasma membrane. One method that has proven to be very helpful to obtain information about the mechanical properties of membranes consists in extracting tethers (nanometer-sized lipid membrane tubes) by the application of a "point" force on the membrane. Tethers were first studied in the early 1970s on red blood cells attached on a glass slide and submitted to shear flows [1]. Various approaches were since then used to extract membrane tethers from cells, but one of the most common and most informative methods is to use micrometer-sized beads trapped by optical tweezers that serve as a sensitive force probe [2–5].

The technique of membrane tether formation and the physics behind it were primarily developed in the study of giant unilamellar vesicles (GUVs) that are formed of pure, usually symmetric lipid

Cedric M. Blouin (ed.), *Caveolae: Methods and Protocols*, Methods in Molecular Biology, vol. 2169,
https://doi.org/10.1007/978-1-0716-0732-9_15,

bilayers. While the principles remain the same when extracting membrane tether from cells, one should keep in mind the corrections and limitations that apply due to the more complex membrane composition and its interaction with the underlying cytoskeleton. In the following, we will briefly discuss how the tether force is related to membrane tension in GUVs and in cells. Further discussion of the limitations and correction factors when extracting membrane tethers from cells will be discussed in the Notes (Subheading 4).

1.1 Physical Description of Membrane Tethers

The key physical parameters of lipid membranes are bending rigidity κ (generally 10–50 k_BT), elastic modulus K (generally 0.2 N/m), and membrane tension σ (generally 10^{-6} to 10^{-2} N/m). K and κ depend on the lipid composition and σ on factors like the pressure difference between interior and exterior of, for example, a GUV or on membrane-cortex interactions in cells (see below). Lipid membranes are basically not extensile, i.e. they rupture upon extension >*3–5%* and their lysis membrane tension is generally $\sigma_{\text{lysis}} = 10^{-2}$ N/m. When a lipid membrane tether (a cylindrical tube of radius R) of length L is pulled, its free energy F_t can be written as

$$F_t = \frac{\pi}{R}\kappa L + 2\pi\sigma RL - fL.$$

Minimizing this with respect to the tether radius leads to

$$R_0 = \sqrt{\frac{\kappa}{2\sigma}}$$

and a tether force at an equilibrium of

$$f_0 = 2\pi\sqrt{2\kappa\sigma}.$$

Note that the tether force is independent of the tether length when the system is at equilibrium.

In the case of cells, the membrane tension σ is generally replaced by an effective membrane tension σ_{eff} to account for the additional effect of the membrane cortex interactions (often represented as interaction energy density W_0) accounting for about 75% of the tether force [6]. The forces to extract and hold cell membrane tethers are generally in the range of 10–50 pN depending on the cell type, the cell stage, and whether the cell is adherent or in suspension.

1.2 Optical Tweezers

The ideal instrument to probe the force range required for pulling lipid membrane tethers is optical tweezers (OT). For experiments on cells, lasers at 1064 nm are most suitable as this wavelength is barely absorbed by water leading to minimal local heating, and lasers with >1 W power are readily available. OT setups are commercially available and manuals on how to build them in various forms are readily available. The two most common approaches are: (1) OT with a static position combined with a movable stage/sample holder or (2) a mobile OT (e.g. via an AOD (acousto optical deflector)) and a static sample. Both approaches are workable for cell tether pulling and please see the notes for further comments. We briefly recall the principle of force measurements with OTs and refer to more detailed literature [7] about the setup of an OT setup as this would go beyond the reach of this chapter.

Generally, we use 1–3 μm diameter polystyrene beads, as they can be easily functionalized to stick to components of the cell plasma membrane. The focused laser beam of the OT forms potential trapping the polystyrene bead. When the bead is displaced by a distance Δx from the center of the OT, the force f_{OT} acting on the bead is

$$f_{\mathrm{OT}} = k_{\mathrm{OT}} P_{\mathrm{laser}}\, \Delta x$$

with k_{OT} representing the stiffness of the OT and P_{laser} being the laser power (Fig. 1a). k_{OT} depends on the difference of refractive indices between the bead and the surrounding medium and the bead diameter, and it is advised to calibrate the OT regularly, especially when the experimental conditions change. Generally, P_{laser} should be chosen to have the effective OT stiffness in the range of $k_{\mathrm{OT}}\ P_{\mathrm{laser}} = 40$–$100$ pN/μm to have a good compromise of force resolution and maximum force that can be applied (*see* **Note 1**).

The following description of tether pulling experiments is based on the experimental setup that was used in [5].

2 Materials

2.1 OT Setup

We worked with a spatially static OT, i.e. the position of the laser trap is fixed. The optical trap was created by steering a 1064 nm laser beam (Coherent, Santa Clara, CA) into an inverted microscope (Axiovert 200, Zeiss), and was directed through a high numerical aperture objective (×100, 1.3 NA) via a hot mirror (Melles Griot) placed under the filter cube holder instead of a slide in analyzer. This design allowed performing bright field or epi-fluorescence imaging while trapping a bead. Analysis of thermal fluctuations of the bead revealed a trap stiffness of $k_{\mathrm{OT},\,x} = 131$ (±6) pN/(μm W). Brightfield images of the bead were acquired with a CCD camera (XCST70CE, Sony, Tokyo, Japan) and analyzed offline using ImageJ (https://imagej.nih.gov/ij/). For life cell experiments, the observation chamber containing cells was mounted on a temperature-controlled microscope stage (Tempcontrol 37-2 digital, Carl Zeiss) set to 37 °C.

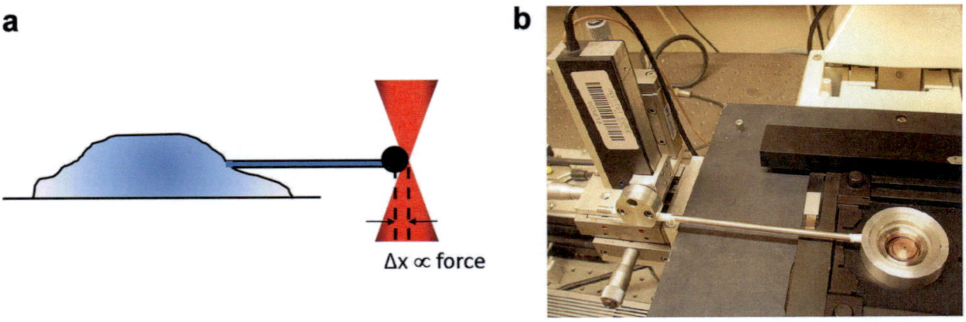

Fig. 1 (**a**) Schematic depicting the relation between the displacement of the bead out of the center of the OT and force during a tether pulling experiment. (**b**) Picture of the cell chamber holder connected to a piezo motor

2.2 Beads

Concanavalin A covered beads (ConA-beads): Incubate 3 μm diameter polystyrene beads in 1 mg/mL Concanavalin A for 30 min at room temperature. Spin down beads for 30 s in a tabletop mini-centrifuge and resuspend in PBS (they can be stored for a month at 4 °C).

2.3 Cell Chamber and Microscope Stage

Since the OT were not movable, cells had to be displaced with respect to the OT to pull tethers with trapped beads. Controlled linear displacement was performed with a piezo-stage (PI, Karlsruhe, Germany). Metal cell chambers (Attofluor, ThermoFisher) were placed in a custom-made holder (shaped like a spoon with a hole) that was connected to the piezo-stage. The aluminum allowed an efficient heat conductance from the temperature control stage to the cells. This allowed us to perform all cell experiments at 37 °C (Fig. 1b).

3 Methods

3.1 Plating of Cells

The day before the planned experiments, cells must be plated on round, 35 mm diameter glass slides.

1. Rinse glass slides five times with EtOH followed by MilliQ water
2. Blow dry glass slides under nitrogen gas stream or by placing them in a dry oven
3. Place 3 glass slides in a 10 cm diameter cell culture dish filled with culture medium
4. Add freshly detached cells at low density (aim 25 cells per 100 μm^2)

3.2 Tether Pulling

Extreme care must be taken when working with OT as the 1064 nm laser is invisible to the human eye. The OT laser is powerful and can cause serious harm to the eyesight and skin burns. Please make sure that appropriate training is provided before handling the setup, to wear appropriate laser protection safety glasses, and that laser interlocks are functional.

1. Preparations
   - (a) Switch on the microscope, laser, and the heating stage (at 37 °C) to allow all system components to warm up (make sure the shutter of the OT is closed)
   - (b) Vortex ConA-bead stock and dilute 1 μL beads in 9 μL PBS
   - (c) Place one glass slide with adherent cells inside on the bottom half of the metal cell chamber, assemble the chamber, and fill with 1–2 mL of pre-warmed cell medium
   - (d) Put oil on the lens of the ×100 microscope objective
   - (e) Place the metal cell chamber into the cell chamber holder such that the chamber is above the objective
   - (f) Approach the objective carefully to the bottom glass of the metal cell chamber until the oil starts to spread and focus your sample using the camera (using eye-pieces elevates the risk of damaging your eyes in case the laser shutter is not closed)
   - (g) Add the diluted ConA-beads (10 μL)
   - (h) Wait for 1 min to allow the beads sink toward the bottom of the chamber
   - (i) Trap a single bead in the OT and move it to a region with a cell and no other beads (*see* **Note 2**)
2. Pulling of a single tether from a cell

   After trapping a bead and finding the cell to perform the tether pull experiment, a typical experiment will consist of the following steps:
   - (a) Position the trapped bead in front of the cell (make sure to be above the surface, i.e. the cell should be slightly out of focus)
   - (b) Start the image acquisition to record the bead position in the absence of a membrane tether (this sets the zero-force reference (*see* **step 3** below) (Fig. 2)
   - (c) Bring the bead into light contact with the cell by moving the cell toward the bead. This pushes the bead out of its reference position (corresponding to the negative force in Fig. 2). Moreover, lowering the z position increases the contact site with the cell membrane.
   - (d) After a few seconds of contact (*see* **Note 3**), move the cell slowly away from the bead (5 μm displacement at ~0.1 μm/s) (*see* **Note 4**).

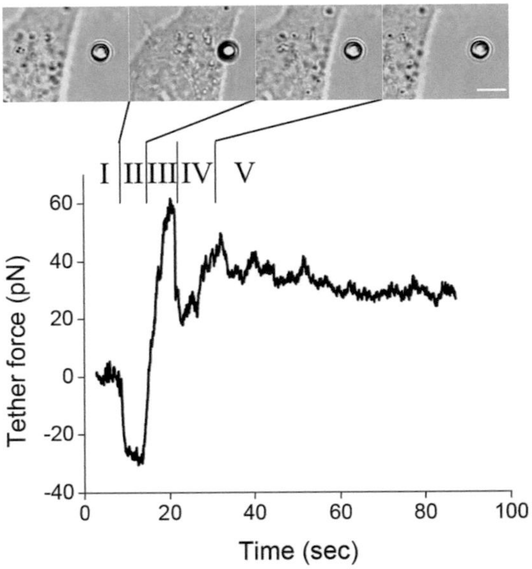

Fig. 2 Typical tether pulling experiment. Top: image sequence of tether formation from an adherent cell; scale bar 5 μm. Bottom: corresponding force trace; the roman figures indicate the different steps of tether extraction: (I) reference force corresponding to the position of the free, untethered bead; (II) negative force corresponding to impingement between bead and cell to establish an adhesive contact; (III) tether nucleation; (IV) tether elongation at pulling velocity $v \sim 0.5$ μm/s (final tether length: 10 μm); (V) tether force relaxation to the equilibrium tether force

(e) An adhesive contact between the bead and the cell membrane will result in a visible displacement of the bead in the direction of the cell (positive force increase in Fig. 2) until a tether is nucleated (force peak) (*see* **Note 5**)

(f) Elongate the tether further at constant velocity (0.5 μm/s), which results in a slow increase of the tether force (*see* **Notes 6–9**).

(g) After $t = 30$ s, stop tether elongation. The tether force should relax to a plateau value, representing the equilibrium tether force, after ~10 s (*see* **Note 10**).

(h) To ensure that a tether was present throughout the experiment, switch off the laser trap to release the bead. The bead should return to the cell if the tether remained intact (*see* **Note 11**).

3. Analysis

Depending on the OT setup, bead position tracking is done optically (thresholding of the bead) or by reading out the signal from a quadrant photodiode. In any case, using the OT calibration, a typical tether pull experiment will result in a force-time trace as depicted in Fig. 2.

To obtain the equilibrium tether force value representing the effective cell membrane tension, one computes the 10 s average value of the recorded force after it relaxed to a plateau (e.g. between 60 and 70 s in Fig. 2).

4 Notes

1. Optical tweezers have a limited potential well in which the force and bead displacement are proportional. If the (3 μm diameter) bead gets displaced out of the trap center by more than 1 μm, the force readout will become unreliable. To avoid this, increase the laser power to keep the bead closer to the trap center. A good test is to pull tethers at different laser powers and to compare the tether forces. If they are similar, the bead stayed in the linear region of the optical tweezers.
2. Trapping beads can be tricky sometimes. Best is to put a marker on the computer screen to indicate the position of the OT. When approaching the bead, keep the OT switched off. Position the OT center slightly above the bead and switch on the laser. Otherwise, it can happen that the bead gets pushed away by the OT.
3. A contact time of 2–20 s is typically sufficient to establish an adhesive link between cell membrane and bead, depending on the cell type, the cell treatment, and the freshness of the bead coating.
4. If the bead escapes the OT, increase the laser power and try again. If this does not help, the attachment time was too long. Trap a new bead and start with a new cell.
5. The force overshoot is the signature of tether formation. The magnitude of the force peak depends on the size of the adhesive contact between membrane and bead [8].
6. In the case of multiple tethers, the force increase will be steeper, and a broader tether or bundle of tethers would become visible.
7. If the bead escapes the OT during tether elongation, usually more than one tether was formed. The bead can be trapped newly with higher laser power to continue the pulling. But it would be better to trap a new bead and pull a tether from another cell.
8. Eventually, the rupture of single tethers with corresponding steps in the tether force would be observed during tether elongation.
9. Elongation of tethers extracted from living cells significantly deviates from the behavior discussed for lipid bilayers

[6, 9]. Instead of exhibiting a length-independent plateau value, the tether force increases upon elongation even at slow pulling velocities (fractions of μm/s). This process originates very likely from viscous friction between cytoskeleton and membrane and from the flow of membrane around transmembrane proteins [10]. A comprehensive physical mechanism of this dynamic contribution to the tether force is still missing and the force velocity relationship is under debate. To avoid the complexity inherent to dynamics of tether extrusion, it is better to focus on the static tether force, as measured after relaxation at a constant length.

10. If the bead shows clear position fluctuations during the relaxation time, actin might have grown into the base of the tether. In this case, the force reading does not represent the effective membrane tension and should be discarded.
11. If actin has grown into the tether, it will remain stiff at its base after the release of the bead.

References

1. Hochmuth RM, Mohandas N, Blackshear JR, Blackshear PL (1973) Measurement of the elastic modulus for red cell membrane using a fluid mechanical technique. Biophys J 13:747–762. https://doi.org/10.1016/S0006-3495(73)86021-7
2. Dai J, Sheetz MP, Wan X, Morris CE (1998) Membrane tension in swelling and shrinking molluscan neurons. J Neurosci 18:6681–6692
3. Cuvelier, D. (2005) Cell adhesion and membrane tubes: some dynamical, mechanical and rheological aspects. PhD thesis, Université Pierre et Marie Curie, Paris VI. http://tel.archives-ouvertes.fr/tel-00010490/en
4. Sheetz MP, Sable JE, Döbereiner H-G (2006) Continuous membrane-cytoskeleton adhesion requires continuous accommodation to lipid and cytoskeleton dynamics. Annu Rev Biophys Biomol Struct 35:417–434. https://doi.org/10.1146/annurev.biophys.35.040405.102017
5. Sinha B, Köster D, Ruez R et al (2011) Cells respond to mechanical stress by rapid disassembly of caveolae. Cell 144:402–413. https://doi.org/10.1016/j.cell.2010.12.031
6. Sheetz MP (2001) Cell control by membrane-cytoskeleton adhesion. Nat Rev Mol Cell Biol 2:392–396. https://doi.org/10.1038/35073095
7. Williams MC (2002) Optical tweezers: Measuring piconewton forces. P. Schwille (Ed.), Single Molecule Techniques, Biophysics Textbook Online. http://www.biophysics.org/Portals/0/BPSAssets/Articles/williams.pdf?ver=2008-07-25-122103-867
8. Koster G, Cacciuto A, Derényi I et al (2005) Force barriers for membrane tube formation. Phys Rev Lett 94:068101. https://doi.org/10.1103/PhysRevLett.94.068101
9. Heinrich V, Leung A, Evans E (2005) Nano- to microscale dynamics of P-selectin detachment from leukocyte interfaces. II. Tether flow terminated by P-selectin dissociation from PSGL-1. Biophys J 88:2299–2308. https://doi.org/10.1529/biophysj.104.051706
10. Campillo C, Sens P, Köster D et al (2013) Unexpected membrane dynamics unveiled by membrane nanotube extrusion. Biophys J 104:1248–1256. https://doi.org/10.1016/j.bpj.2013.01.051

Chapter 16

Live Confocal Imaging of Zebrafish Notochord Cells Under Mechanical Stress In Vivo

Ye-Wheen Lim, Harriet P. Lo, Thomas E. Hall, and Robert G. Parton

Abstract

The zebrafish is a vertebrate model suited to the exploration of cell biology within a whole organism. Hypotheses in cell mechanics can be tested by using the zebrafish notochord as a manipulable experimental system. Here, the methodologies to prepare, label, and simultaneously induce and image mechanical loading on live zebrafish notochord cells via electrical stimulation are described. This approach investigates membrane mechanics in a live, physiological setting and is thus suited for caveola research where observations within the tissues of an intact organism are increasingly relevant. This chapter also aims to introduce fundamental methodologies for the use of zebrafish in "in vivo cell biology."

Key words Zebrafish, Notochord, Live imaging, In vivo, Mechanical stress, Electrical stimulation, Confocal microscopy, Caveolae, Cavin, Vertebrate

1 Introduction

A comprehensive understanding of caveola biology will increasingly necessitate the use of *in vivo* systems. Caveolae and their protein components (Cavins and Caveolins) are involved in a variety of physiological and disease-related roles covering many distinct cell types including the notochord, skeletal muscles, and adipose [1–3]. The small, optically transparent, and genetically manipulable zebrafish is a vertebrate model amenable to live experiments, complementing conventional mammalian systems such as mice and rats, and simpler systems such as cells in culture. The distribution and development of caveolae within zebrafish tissues has been well characterized [4–9]. This, in addition to their optical transparency in early development, provides a unique opportunity to probe caveola-related biology via live imaging, where any tissue or cell type of interest can be examined with a high degree of spatial and temporal resolution.

A growing number of studies pertaining to the roles of caveolae in relation to membrane mechanical stress have emerged [10–14],

Cedric M. Blouin (ed.), *Caveolae: Methods and Protocols*, Methods in Molecular Biology, vol. 2169,
https://doi.org/10.1007/978-1-0716-0732-9_16,

bringing about *in vivo* research in the area of caveola mechanoprotection [8, 15, 16]. In particular, we and others have developed methods to utilize the zebrafish notochord as a system to study the function of caveolae in response to mechanical stress using caveola-deficient cells [17–19]. The notochord, an evolutionarily conserved hydrostatic skeleton, is composed of inner notochord cells enveloped by epithelial-like sheath cells, which are juxtaposed against an outer perinotochordal sheath of extracellular matrix [5, 20]. Notochord cells possess fluid-filled intracellular vacuoles which provide flexural stiffness and mechanical strength to the structure via hydrostatic pressure [21]. As a hydrostatic skeleton, the cylindrical notochord is intrinsically analogous to a structural beam susceptible to bending moments. In the zebrafish, the embryonic notochord is an optically visible rod surrounded by skeletal muscles, which modulate its flexion. By applying an external electrical field across the trunk region of a zebrafish embryo, action potential mediated by distinct voltage-gated ion channels in neurons and skeletal muscle fibers can be triggered, leading to muscle contraction [17, 22, 23]. Thus, electrical stimulation is able to increase mechanical loading on the notochord by inducing lateral bending and axial contraction of the notochord via contraction of adjoining skeletal muscles [22].

Here, we describe time series confocal imaging of fluorophore-labeled zebrafish notochord cells under mechanical stress. The steps to prepare, label, mount, and simultaneously stimulate and image a zebrafish embryo are detailed. We have shown that when appropriately applied, electrical stimulation does not notably affect the survival of stimulated animals or the morphology of wild-type (WT) zebrafish skeletal muscles [17]. These methodologies can be used for quantitative and qualitative comparison of different experimental groups, such as gain or loss of function assays in relation to mechanical stress in live whole organisms.

2 Materials

2.1 Zebrafish

1. Zebrafish line(s) of interest (*see* **Note 1**). Zebrafish WT line with the *TAB* background is used.
2. Zebrafish are raised, maintained, and bred according to institutional guidelines (*see* **Note 2**). In general, adult zebrafish are housed in 3 or 8 L tanks with flow and embryos up to 7 days postfertilization (dpf) are incubated in the dark in 10 cm plastic dishes with E3 media. Temperature is maintained at 28.5 °C.

2.2 Reagents and Buffers

1. E3: Working concentration at 5 mM NaCl, 0.17 mM KCl, 0.33 mM $CaCl_2$, and 0.33 mM $MgSO_4$ in ultrapure water. 10 L of 60× stock solution is prepared by mixing 172 g

NaCl, 7.6 g KCl, 29 g $CaCl_2.2H_2O$, and 49 g $MgSO_4.7H_2O$ with ultrapure water. 60× stock solution is stored at 4 °C.

2. E3/1-phenyl-2-thiourea (PTU): 100 mM PTU stock is prepared by adding 0.152 g PTU in 10 mL of ethanol. 5 mM PTU stock is prepared by diluting 1 mL of 100 mM PTU stock in 19 mL of E3. Stocks are stored in a fume hood at room temperature.
3. Ethyl 3-aminobenzoate methanesulfonate (tricaine): 25× stock solution is prepared by adding 400 mg tricaine powder to 98 mL of ultrapure water. Approximately 2 mL of 1 M Tris, pH 9 is then added. Ensure stock solution pH is at 7. Store in the dark at 4 °C for up to a week or long-term at −20 °C.
4. BODIPY FL C5-Ceramide (ThermoFisher Scientific): 1 mg/mL stock solution is prepared by dissolving 250 μg solid BODIPY FL C5-Ceramide in 416 μL of dimethyl sulfoxide (DMSO) and stored at −20 °C.
5. Low gelling temperature (LGT) agarose solution (Sigma-Aldrich): 0.8% LGT agarose solution is prepared first by dissolving the agarose mixture (0.24 g LGT agarose in 30 mL of E3) completely in the microwave. Allow the solution to cool for 1–2 min at room temperature. Mix the solution by swirling and pipette 1 mL aliquots into 1.5 mL Eppendorf tubes for storage at −20 °C.
6. Pronase: 30 mg/mL stock solution is prepared by adding 1 g of lyophilized pronase to 33.33 mL of E3 and stored at −20 °C.

2.3 Equipment, Tools, and Embedding Components

1. Imaging plate. A flexible silicone imaging mold is created using a microinjection mold (Adaptive Science Tools) set against Sylgard 184 Silicone Elastomer components as per manufacturer's instructions. The microinjection mold was pre-sprayed with Stoner A324 Rapid Release. The mold is then roughly cut into a 6.5 cm × 3.7 cm piece and placed into a 10 cm dish (Fig. 1a, b).
2. 35 mm plastic dish.
3. 10 cm plastic dish.
4. Micropipettes.
5. Plastic Pasteur pipettes.
6. Plastic Pasteur pipettes with fine tip.
7. Stereomicroscope (conventional, such as Olympus SZX7).
8. Constant-temperature heating incubator (any conventional ones that accommodates 1.5 mL tubes such as the Eppendorf Thermomixer 5436).
9. Dry bath incubator heat block (Thermo Fisher Scientific).
10. 24-well plate.

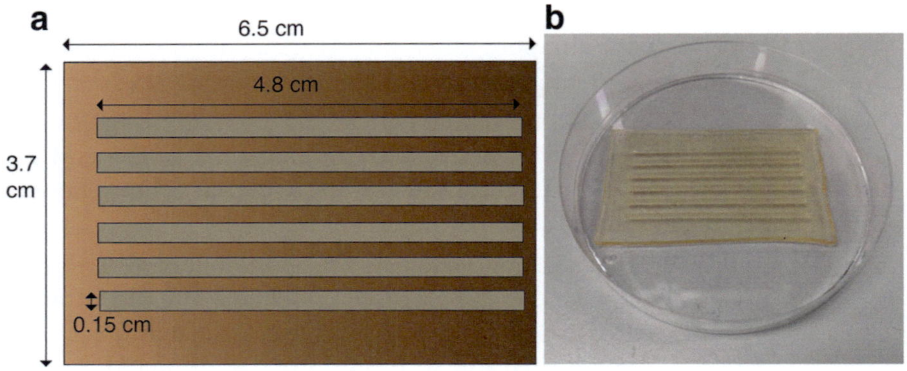

Fig. 1 (**a**) Dimensions of the cut silicon mold. (**b**) Imaging plate consists of the cut silicon mold placed in a 10 cm plastic dish

Fig. 2 (**a**) Dimensions of the plastic ring piece. (**b**) Electrode ring attached to a representative confocal microscope objective. (**c**) Hook electrodes are insulated in PVC and connector wires are held in place by the outer O-ring

11. Tools to orient zebrafish embryos. Blunt forceps, fine pipette tips or Microloader tips cut to a length of 15 mm are used.
12. Zebrafish incubator (conventional, such as Thermoline Scientific TI-500F).
13. Sharp forceps such as Dumont no. 5 forceps.

2.4 Electrical Stimulator Component

1. Electrical stimulator. S44 Square Pulse Stimulator, constant voltage (Grass Instruments).
2. SIU5 RF Transformer Isolation Unit (Grass Instruments).
3. Electrode ring. A ring piece made out of a section of a plastic tube plastic tube is recessed with two grooves on the outer and inner sides for O-rings to be able to fit on both sides (Fig. 2a). The outer O-ring sits on the outer groove and holds platinum

electrode wires underneath it. The wires are insulated with PVC. The inner O-ring is fitted in the inner groove and is necessary to hold the electrode ring against a confocal microscope objective (Fig. 2b). The rings have been drilled so that platinum electrode wires run in from one side of the electrode ring and underneath the inner O-ring (Fig. 2c).

2.5 Microscope Components

1. Zeiss LSM 710 upright confocal microscope. ZEN 2012 (black edition) microscopy software installed.
2. 40×/1.0 W Plan-Apochromat DIC M27 water immersion objective.

3 Methods

3.1 Sample Preparation

Two dpf zebrafish embryos are used during the imaging step. The developmental stage of embryos used must be similar to ensure consistency between experimental groups (*see* **Note 3**). Two dpf embryos are staged as being in the high pec to long pec developmental stages by observing the presence of emerging pectoral fin bud. At approximately 1 dpf, embryos are prepared for imaging and treated for both pigmentation (*see* **Note 4**) and chorion (*see* **Note 5**) via the following procedure:

1. Fresh E3/PTU working solution is prepared by adding 1 mL of 5 mM PTU stock solution to 30 mL E3.
2. At 21–24 h post fertilization (hpf), between 26-somite and prim-5, and before pigments can be observed (*see* **Note 6**), transfer zebrafish embryos using a plastic Pasteur pipette (*see* **Note 7**) into a 10 cm dish with fresh E3/PTU working solution. From this point in the methodology, E3/PTU replaces E3 as embryo media.
3. Add approximately 100 μL of pronase stock solution into the 10 cm plastic dish containing 30 mL of E3/PTU and embryos, diluting to a working pronase concentration of 0.1 mg/mL.
4. Swirl the dish and allow for 40 min rest at 28.5 °C.
5. Using a plastic Pasteur pipette, gently pipette the embryos and release them back into the media at least 2 times. Floating, separated chorions should be visible.
6. Wash 3 times with E3/PTU by gently pouring the media and floating chorions out into a waste container. The embryos should remain on the surface of the dish during the washes.

3.2 Zebrafish Labeling

After dechorionation, 1 dpf embryos are ready for labeling. This process can be carried out immediately after or later in the day as labeling occurs overnight. Fluorescent sphingolipid BODIPY FL C5-Ceramide is used as a transient vital stain to delineate the plasma membrane. The following labeling method is relatively simple and eliminates the need for a transgenic zebrafish line in this imaging process:

1. Using a plastic Pasteur pipette, transfer 7 to 10 zebrafish embryos from the 10 cm plastic dish to a 24-well plate, ensuring that a total of 1 mL of E3/PTU per well is transferred along with embryos (*see* **Note 8**).
2. Add 5 μL of BODIPY FL C5-Ceramide stock solution to each well containing 1 mL of transferred E3/PTU and embryos.
3. Mix by pipetting gently until color of media appears homogenous.
4. Incubate at 28.5 °C overnight in the zebrafish incubator (*see* **Note 9**).
5. Wash BODIPY FL C5-Ceramide labeled embryos for 1 min 3 times by gently transferring embryos into fresh E3/PTU each time, either in new 35 mm dishes or in a different well on the same 24-well plate. Perform this step immediately before the next mounting step.

3.3 Zebrafish Mounting

At 2 dpf, labeled embryos are anesthetized and mounted in an imaging plate in preparation for electrical stimulation and confocal microscopy.

1. Prepare tricaine working solution by adding 1 mL of stock tricaine solution to 24 mL of E3/PTU to make tricaine/E3/PTU media (*see* **Note 10**).
2. Anesthetize embryos by transferring washed, BODIPY FL C5-Ceramide labeled embryos into tricaine working solution, either in new 35 mm dishes or in a different well on the same 24-well plate.
3. Lightly touch the tail of random embryos with a fine tip Pasteur pipette to verify inactivity.
4. Thaw stock 0.8% LGT agarose solution tube at 95 °C uncovered for 1 min in a dry bath incubator heat block, mix the tube by inverting and keep the melted 0.8% LGT agarose solution tube at 37 °C in an Eppendorf Thermomixer 5436 incubator.
5. Gently pipette 1 to 3 embryos into the 0.8% LGT agarose solution tube. Immediately and gently, pipette the embryos and approximately 200 μL of 0.8% LGT agarose solution from the tube into an imaging mold groove of the imaging plate.
6. Using a fine tip plastic Pasteur pipette, remove excess overflowing agarose solution in the imaging mold groove.
7. Under a stereomicroscope, immediately orient the embryos in a consistent position (*see* **Note 11**). Anterior to the left, posterior to the right, and lateral side parallel to plate surface.

8. Allow gel to cool and solidify for 5 min (*see* **Note 12**).
9. Fill the imaging plate with tricaine/E3/PTU media (approximately 25 mL). Ensure the imaging mold and embryos are completely immersed.
10. Incubate at microscopy room temperature (*see* **Note 13**) for at least 30 min (*see* **Note 14**).

3.4 Electrical Stimulation Setup

Prepare embryos for simultaneous electrical stimulation and live imaging via the following:

1. Connect the Grass SIU5 stimulus isolation unit to the Grass S44 electrical stimulator output.
2. Fit the electrode ring on the 40×/1.0 W Plan-Apochromat DIC M27 objective carefully (Fig. 2b) and connect electrode leads to the SIU5 output.
3. Use the following settings for the Grass S44 electrical stimulator: 0.04 pulse per second (pps) stimulation rate, 0.9 ms delay, 0.8 ms duration, and 11.6 V voltage (*see* **Note 15**).

3.5 Confocal Microscopy, Image Acquisition, and Analysis

Confocal imaging is carried out on BODIPY FL C5-ceramide labeled zebrafish embryos at the developmental stage of 2 dpf (Fig. 3). Different confocal setups with access to a high-resolution water immersion objective can be used. In this method, the Zeiss LSM 710 upright confocal microscope is used alongside the ZEN 2012 software to acquire images. The following are general steps to follow:

1. Perform computer and confocal microscope startup, including installation of the 40×/1.0 W Plan-Apochromat DIC M27 water immersion objective if necessary.
2. Ensure that the imaging plate containing embryos is secured flat and seated firmly on the microscope stage.
3. Gently immerse the water objective into the imaging plate (*see* **Note 16**).
4. Locate the notochord at the midline of the embryo and choose a region of interest (*see* **Note 17**).
5. Choose appropriate filters for the label of interest, which in this case is for BODIPY FL.
6. Optimize image acquisition time by adjusting acquisition parameters resulting in a scan time of 7.75 s (frame size 512 × 512 pixels, speed 7, averaging 2). Verify that the cell or tissue of interest is in focus and all relevant information can be captured. Stagger the acquisition time in relation to the electrical stimulator stimulation rate to ensure that images are not acquired during a twitch (*see* **Note 18**). A time series with a duration of 50 cycles and an interval of 92.3 s is set up (*see* **Note 19**).

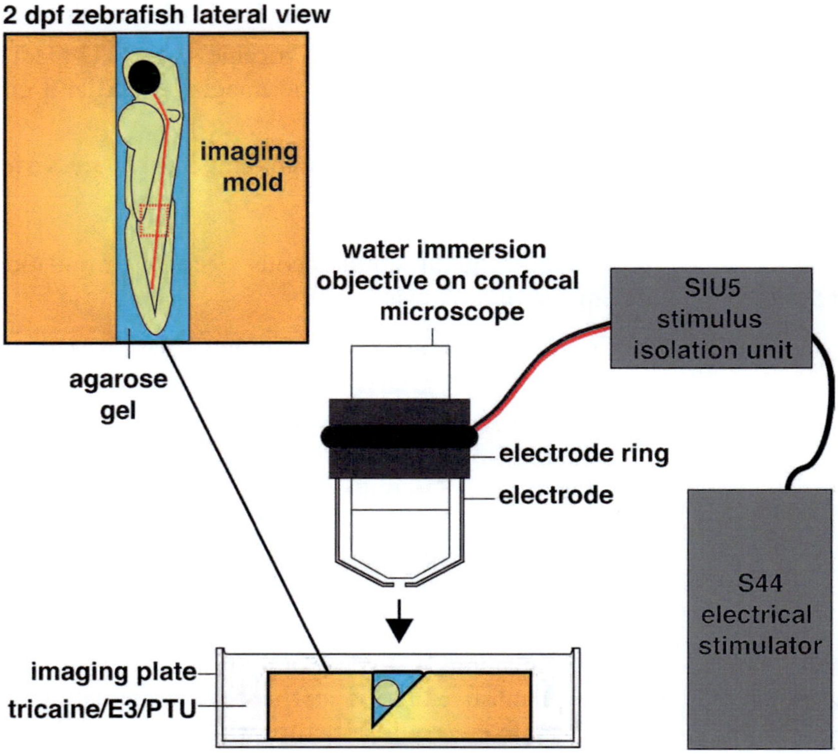

Fig. 3 Schematic illustration displaying the imaging process. BODIPY FL C5-Ceramide labeled 2 dpf zebrafish embryo is immobilized in tricaine/E3/PTU media and embedded in LGT 0.8% agarose gel. Water immersion objective is dipped into tricaine/E3/PTU (arrow), placing electrodes from the electrode ring directly over the zebrafish trunk. Electrodes are connected to the S44 electrical stimulator via the SIU5 stimulus isolation unit. Imaging is carried out on the confocal microscope while the embryo is being stimulated. Lateral view inset: zebrafish embryo orientation is anterior to the top, posterior to the bottom. Notochord in red. Dotted red box is a representative region of interest

7. Switch on the power to the Grass S44 electrical stimulator.
8. Turn on the output on the Grass S44 electrical stimulator and start the time series experiment at the same time to stagger the period between electrical pulses and image acquisition.
9. Use image analysis software such as ImageJ/Fiji to analyse acquired time series data (Fig. 4a–c).
10. If recovery of stimulated animals is required, gently break apart the agarose gel surrounding an embryo of interest with sharp forceps under a stereomicroscope. Use a plastic Pasteur pipette to gently transfer the embryo into a new dish with fresh E3 or E3/PTU media (some agarose gel can still be attached to the embryo). Embryos should begin to swim and break out of the residual gel.

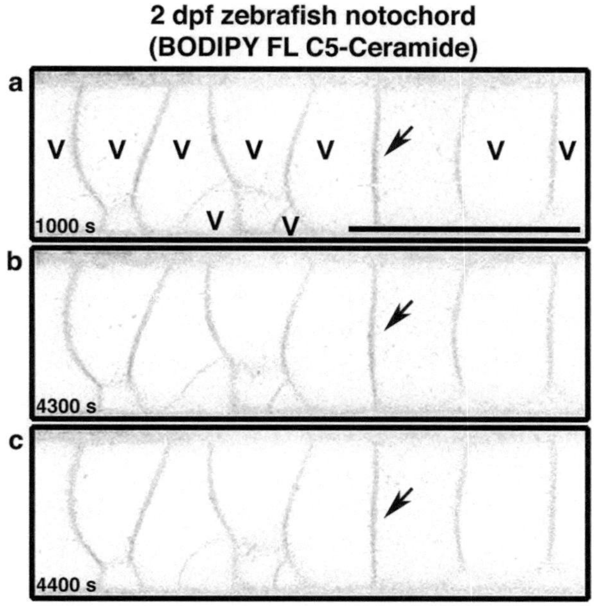

Fig. 4 Selected time series images of a BODIPY FL C5-Ceramide labeled WT zebrafish notochord under electrical stimulation (100 s time series interval; shown as inverted color images). (**a–c**) The notochord under electrical stimulation at 1000 s and 10 cycles, 4300 s and 43 cycles, and 4400 and 44 cycles respectively. At this point the notochord had experienced 40, 172, and 176 pulses of stimulation respectively. Each vacuole (v) indicates one notochord cell. WT zebrafish notochord cells do not appear to possess prominent morphological changes besides slight arching of delineated membranes (arrows). Scale bar = 100 μm. Embryo orientation: anterior to the left, posterior to the right

4 Notes

1. The *TAB* line is an *AB/TU* line generated by The University of Queensland Biological Resources Aquatics Facility. Zebrafish lines with the *TAB* background were used to optimize the methods described here. However, these methods should be applicable to most zebrafish lines of interest (such as CRISPR/Cas9 mutants) with a phenotype that permits consistent mounting and imaging. The generation of genome-edited lines using CRISPR/Cas9 has already been described in detail in the literature [1, 2].
2. Zebrafish husbandry is generally based on The Zebrafish Book [3].
3. Depending on the phenotypes of lines used, the timing of embryo production (such as time differences in laying of breeding pairs, removal of dividers, and egg collection) and general environment of clutches (such as fluctuations in temperature),

embryos obtained may reach different developmental stages at a specific time point. Stage the embryos at different time points before 2 dpf to ensure developmental similarity [4]. For example, at 5 hpf, 12 hpf, and 1 dpf.

4. PTU treatment is performed to inhibit pigment formation which optically obstructs the notochord and introduces fluorescence imaging artifacts.
5. Dechorionation at 1 dpf is carried out to avoid variation in mechanical stimulus experienced, ensuring a similar period of hatching and thus notochord flexion. Additionally, environmental variations in **Note 3** may also result in a temporal difference in embryo hatching, resulting in unhatched embryos at the desired imaging timepoint of 2 dpf.
6. It is vital to perform PTU treatment before pigment formation as PTU does not inhibit or reduce existing pigments. PTU can be added as early as after the gastrula period (approximately 10 hpf).
7. Use different plastic pipettes for different experimental groups (such as different lines or clutches). Zebrafish embryos, especially after dechorionation can stick to the inner walls of plastic pipettes.
8. To prevent reagent wastage, a total of 1 mL per well is used in a 24-well plate. The dimensions of each well in a 24-well plate will provide sufficient water level height and horizontal distance for embryos to thrive in the reagent. Depending on experimental design, each well can accommodate a different experimental group, such as a different clutch or zebrafish line.
9. 8 h minimum incubation period is needed for pronounced notochord labeling.
10. Always make up fresh tricaine working solution, as aged 1× tricaine solution (particularly after being left at room temperature for 1 day) can be toxic to embryos.
11. After approximately 1 min, depending on room temperature, as the agarose gel solidifies, it becomes increasingly difficult to orient the embryos without damaging them. Thus, it is vital to practice orienting embryos with a comfortable orientation tool.
12. The cooling period can vary based on room temperature and agarose solidification can be difficult to tell by eye. After embryo orientation, a drop of 0.8% agarose solution can be pipetted into a new dish as a comparison standard. This agarose drop can be checked by eye or forceps over time to monitor and compare the solidification period.

13. In this chapter, imaging was performed at room temperature as our upright confocal microscopy setup does not possess a heating stage or incubator. In general, zebrafish experiments should be performed at 28.5 °C and temperature changes should be noted.
14. Incubation of at least 30 min to 1 h is needed to acclimatize and stabilize fresh agarose gel at the desired working temperature. Variation in gel stability can result in focus drift during timelapse imaging.
15. It is recommended to first optimize electrical stimulator settings under a stereomicroscope. In our case, the electrode ring was taped to the objective. Note that electrode placement in relation to the trunk will affect the resulting electric field strength on muscle contraction. This is one factor that we could not account for when optimizing under a stereomicroscope. Ensure that depth of the electrode ring on the confocal objective and microscope stage height are consistent. When optimizing stimulator settings, look for a balance between voltage and stimulation rate. Here, a supramaximal voltage of 11.6 V was used due to a low stimulation rate of 0.04 pps. To determine our voltage setting, we first assessed the contraction threshold for our WT zebrafish line, which is the minimum voltage required to observe a twitch response at 1 pps. Voltage is then gradually increased as we determined our imaging period and decreased our stimulation rate (covered in **Note 18**). Under the stereomicroscope, our *cavin1b*$^{-/-uq7rp}$ zebrafish line with caveola-deficient notochord cells was also able to act as a positive control for a consistent readout to different stimulation settings, before performing confocal imaging [5]. Whenever electrical stimulation is being carried out, even after optimization, frequently ensure that embryos look morphologically healthy, possess normal heart rhythm, and consistent blood flow can be observed in the vasculature, which is visible with the notochord as a region of interest.
16. Ensure that adequate water level is maintained in the imaging plate throughout the imaging process. This will prevent focus drift and embryos from drying out. Evaporation can be reduced by taping the remaining opening of the plate or making a custom plate cover that allows the objective to be immersed. The water level can also be maintained by routinely adding tricaine/E3/PTU into the plate gently via a plastic pipette.
17. If a constant region between embryos is desired, a standard such as the tip of the yolk extension can be used from embryo to embryo (refer to Fig. 3 inset). Additionally, somatic boundaries can be used as a reference or "scale" to be compared against surrounding tissues.

18. A guideline to staggering the time series and avoiding image acquisition during muscle contraction is to verify that the acquisition time is below the stimulation rate interval period. Additionally, ensure that the total acquisition time plus time series interval period is a multiple of the pulse interval. Be aware of phototoxic and photobleaching effects of long acquisition times. For the time series in this chapter, as the stimulation rate is 0.04 pps (1 pulse every 25 s), image is acquired with a scan time of 7.75 s, at an interval of 92.3 s.
19. With our electrical stimulator settings, zebrafish embryos managed to persist and survive after a stimulation period of at least 156 min.

Acknowledgements

We acknowledge Yeping Wu for helpful discussions during the optimization process. This work was supported by fellowships and grants from the National Health and Medical Research Council of Australia (to R.G. Parton, grant numbers 569542 and 1045092, to T.E. Hall and R.G. Parton, grant number APP1099251) as well as by the Australian Research Council Centre of Excellence in Convergent Bio-Nano Science and Technology to R.G. Parton. Confocal microscopy was carried out at the Australian Cancer Research Foundation (ACRF)/Institute for Molecular Bioscience (IMB) Dynamic Imaging Facility for Cancer Biology, established with funding from the ACRF.

References

1. Lamaze C, Torrino S (2015) Caveolae and cancer: a new mechanical perspective. Biom J 38:367–379
2. Parton RG (2018) Caveolae: structure, function, and relationship to disease. Annu Rev Cell Dev Biol 34:111–136
3. Woodman SE, Sotgia F, Galbiati F, Minetti C, Lisanti MP (2004) Caveolinopathies: mutations in caveolin-3 cause four distinct autosomal dominant muscle diseases. Neurology 62:538–543
4. Flynn EJ 3rd, Trent CM, Rawls JF (2009) Ontogeny and nutritional control of adipogenesis in zebrafish (Danio rerio). J Lipid Res 50:1641–1652
5. Nixon SJ, Carter A, Wegner J, Ferguson C, Floetenmeyer M, Riches J, Key B, Westerfield M, Parton RG (2007) Caveolin-1 is required for lateral line neuromast and notochord development. J Cell Sci 120:2151–2161
6. Nixon SJ, Wegner J, Ferguson C, Mery PF, Hancock JF, Currie PD, Key B, Westerfield M, Parton RG (2005) Zebrafish as a model for caveolin-associated muscle disease; caveolin-3 is required for myofibril organization and muscle cell patterning. Hum Mol Genet 14:1727–1743
7. Hill MM, Bastiani M, Luetterforst R, Kirkham M, Kirkham A, Nixon SJ, Walser P, Abankwa D, Oorschot VM, Martin S, Hancock JF, Parton RG (2008) PTRF-Cavin, a conserved cytoplasmic protein required for caveola formation and function. Cell 132:113–124
8. Lo HP, Nixon SJ, Hall TE, Cowling BS, Ferguson C, Morgan GP, Schieber NL, Fernandez-Rojo MA, Bastiani M, Floetenmeyer M, Martel N, Laporte J, Pilch PF, Parton RG (2015) The caveolin-cavin system plays a conserved and critical role in mechanoprotection of skeletal muscle. J Cell Biol 210:833–849

9. Fang PK, Solomon KR, Zhuang L, Qi M, McKee M, Freeman MR, Yelick PC (2006) Caveolin-1alpha and -1beta perform nonredundant roles in early vertebrate development. Am J Pathol 169:2209–2222
10. Cheng JPX, Nichols BJ (2016) Caveolae: one function or many? Trends Cell Biol 26:177–189
11. Nassoy P, Lamaze C (2012) Stressing caveolae new role in cell mechanics. Trends Cell Biol 22:381–389
12. Echarri A, Del Pozo MA (2015) Caveolae - mechanosensitive membrane invaginations linked to actin filaments. J Cell Sci 128:2747–2758
13. Parton RG, del Pozo MA (2013) Caveolae as plasma membrane sensors, protectors and organizers. Nat Rev Mol Cell Biol 14:98–112
14. Sinha B, Koster D, Ruez R, Gonnord P, Bastiani M, Abankwa D, Stan RV, Butler-Browne G, Vedie B, Johannes L, Morone N, Parton RG, Raposo G, Sens P, Lamaze C, Nassoy P (2011) Cells respond to mechanical stress by rapid disassembly of caveolae. Cell 144:402–413
15. Cheng JP, Mendoza-Topaz C, Howard G, Chadwick J, Shvets E, Cowburn AS, Dunmore BJ, Crosby A, Morrell NW, Nichols BJ (2015) Caveolae protect endothelial cells from membrane rupture during increased cardiac output. J Cell Biol 211:53–61
16. Elliott MH, Ashpole NE, Gu X, Herrnberger L, McClellan ME, Griffith GL, Reagan AM, Boyce TM, Tanito M, Tamm ER, Stamer WD (2016) Caveolin-1 modulates intraocular pressure: implications for caveolae mechanoprotection in glaucoma. Sci Rep 6:37127
17. Lim Y-W, Lo HP, Ferguson C, Martel N, Giacomotto J, Gomez GA, Yap AS, Hall TE, Parton RG (2017) Caveolae protect notochord cells against catastrophic mechanical failure during development. Curr Biol 27:1968–1981.e1967
18. Garcia J, Bagwell J, Njaine B, Norman J, Levic DS, Wopat S, Miller SE, Liu X, Locasale JW, Stainier DYR, Bagnat M (2017) Sheath cell invasion and trans-differentiation repair mechanical damage caused by loss of caveolae in the zebrafish notochord. Curr Biol 27:1982–1989
19. Tillu VA, Lim YW, Kovtun O, Mureev S, Ferguson C, Bastiani M, McMahon KA, Lo HP, Hall TE, Alexandrov K, Collins BM, Parton RG (2018) A variable undecad repeat domain in cavin1 regulates caveola formation and stability. EMBO Rep 19:e45775
20. Parsons MJ, Pollard SM, Saude L, Feldman B, Coutinho P, Hirst EM, Stemple DL (2002) Zebrafish mutants identify an essential role for laminins in notochord formation. Development 129:3137–3146
21. Ellis K, Hoffman BD, Bagnat M (2013) The vacuole within: how cellular organization dictates notochord function. BioArchitecture 3:64–68
22. Attili S, Hughes SM (2014) Anaesthetic tricaine acts preferentially on neural voltage-gated sodium channels and fails to block directly evoked muscle contraction. PLoS One 9:e103751
23. Martin BL, Gallagher TL, Rastogi N, Davis JP, Beattie CE, Amacher SL, Janssen PML (2015) In vivo assessment of contractile strength distinguishes differential gene function in skeletal muscle of zebrafish larvae. J Appl Physiol 119:799–806

Chapter 17

Study of Caveolae-Dependent Mechanoprotection in Human Muscle Cells Using Micropatterning and Live-Cell Microscopy

Melissa Dewulf and Cedric M. Blouin

Abstract

Caveolae are plasma membrane organelles that are, among many other features, involved in mechanosensing and mechanoprotection. Different tools have been developed to study caveolae-dependent mechanoprotection and had to be adapted to the tissue or cells studied, as these structures are found in almost every type of cells. This chapter focuses on a protocol combining the use of live-cell imaging, micropatterning, hypo-osmotic shock as a mechanical stress, and dyes such as calcein-AM and propidium iodide. We used this protocol for the in vitro study of the effect of mechanical stress on membrane integrity in human muscle cells from patients bearing caveolin-3 mutations.

Key words Caveolae, Mechanoprotection, Micropatterning, Live-cell imaging, Muscle cells

1 Introduction

Caveolae are cup-shaped membrane invaginations that are present in almost every cell types but are particularly enriched in cells experiencing mechanical constraints such as adipocytes, endothelial and muscle cells. Caveolae were first associated with cell trafficking, lipid homeostasis, and cell signaling but more recently, they have been described to play a role in cell mechanosensing and mechanoprotection [1, 2]. They can be found alone, or interconnected to form structures called rosettes, and have the capacity to flatten out in cells undergoing mechanical stress. This flattening provides additional membrane that allows preventing membrane damage by buffering PM tension increase [3, 4]. This essential property has been confirmed in different in vitro and in vivo studies performed on either cell lines or genetically transformed animals knock out or mutated for different key components of caveolae such as caveolin-1 (Cav1), caveolin-3 (Cav3), or cavin-1 [4–8]. Up to now, many different tools have been used to study the role of caveolae in

Cedric M. Blouin (ed.), *Caveolae: Methods and Protocols*, Methods in Molecular Biology, vol. 2169,
https://doi.org/10.1007/978-1-0716-0732-9_17,

mechanoprotection, with specific adaptation regarding the models, tissues, or cells that are studied. For example, CAV1 KO mice challenged with either an acute increase of cardiac output or hypoxia were used to study caveolar mechanoprotection in endothelial cells and arteries [8]. Swimming, movement challenge within viscous medium, or forced muscle contraction were used on zebrafish, mutated for caveolae genes, to study their function in notochord or the muscle integrity [4, 6, 7]. To study membrane damage, tissue samples from mechanically challenged animals have been observed by electron microscopy and immunofluorescence to carefully study morphological changes and membrane lesions. In vitro, Cav3 mutant muscle cells and CAV1 KO fibroblasts were challenged with a hypo-osmotic shock [3, 5]. Fluorescent or contrasting agents can also be used in the animals to look at infiltration of the dyes as a consequence of cell membrane rupture in different tissues [4, 6–8]. For in vitro studies, the use of life-dead kit allowed determining the proportion of dead cells after mechanical stress, which indirectly gives the amount of cells with membrane damage without any temporal information [5]. To study directly membrane damage, one assay is commonly used and will be the focus of this book chapter. In endothelial cells, the calcein-AM dye has been used in combination with ethidium homodimer-1 to monitor membrane damage after a hypo-osmotic shock [5]. Calcein-AM allows visualizing intact cells whereas ethidium homodimer-1 indicates damaged cells, as it can go only in ruptured cells and bind their DNA. For our studies, we have adapted this technique to investigate mechanoprotection in human muscle cells from patients bearing caveolin-3 mutations [9]. We combined the use of calcein-AM and propidium iodide (PI) with micropatterning and live-cell imaging. Myotubes were aligned on line patterns to be closer to physiological conditions and facilitate image analysis. Moreover, by adding live-cell imaging, we then have access to the temporal study of cell membrane integrity upon mechanical stress in muscle cells from patients. Thus, in addition to the mean percentage of damaged cells, we can determine the mean time of burst.

Here we describe a protocol designed for the study of human myotubes, but this technique can be adapted to study all kinds of adherent cells, using or not micropatterns. For example, we also used successfully this assay with human melanocytes and HeLa cells.

2 Materials

2.1 Cell Culture

Human immortalized myoblasts are grown at 37 °C, 5% CO_2 in Skeletal Muscle Cell Growth Medium (Promocell) supplemented with 20% Fetal Calf Serum (FCS), 50 μg/mL of fetuine, 10 ng/mL of epidermal growth factor, 1 ng/mL basic fibroblast growth factor, 10 μg/mL of insulin, and 0.4 μg/mL of dexamethasone. Prior to experiments, myoblasts are seeded on the surface coated with 1% matrigel (v/v) diluted in PBS at 37 °C for 15 min. To reach 80–100% confluency on line micropatterns such as described below, the amount of cells normally used for a total surface should be divided by 3.

For myoblast differentiation into myotubes, confluent myoblasts are put in DMEM high-glucose Glutamax, supplemented with 0.1% of insulin (v/v) for 4 days.

2.2 Micropatterning

1. PLL-g-PEG (PLL(20)-g(3.5)-PEG [2]): 1 mg/mL in 10 mM Hepes pH 8.6. Adjust pH to 7.4 with NaOH solution. It is a 10× solution. Store at 4 °C.
2. 18 mm glass coverslips.
3. UV ozone oven 185 nm equipped with an ozone catalyzer (UVO cleaner, model 42, Jelight) for deep UV illumination.
4. Synthetic Quartz mask with lines of 10 μm of width, separated by 60 μm (Delta Mask, Toppan, Selba Tech).
5. Ethanol 70%, double distilled and autoclaved water (ddH_2O) and PBS at room temperature (RT).

2.3 Hypo-osmotic Shock and Live-Cell Microscopy

1. Calcein-AM (Life Technologies): 1 mg/mL in dimethyl sulfoxide (DMSO) (*see* **Note 1**). This is a 200× solution. Store at −20 °C in the dark.
2. DAPI: 5 mg/mL in PBS. This is a 100× solution. Store at −20 °C in the dark.
3. Propidium iodide (PI): 10 mg/mL in ddH_2O. This is a 5× solution. When dissolved in hypo-osmotic shock medium, PI volume needs to be considered compared to the total volume. Store at −20 °C in the dark.
4. Hypo-osmotic shock medium: 10% (v/v) differentiation medium in ddH_2O, 30 mOsm final supplemented with 2 mg/mL PI (1×) solution. Prepare fresh prior to the experiment.
5. Matrigel: 1% in PBS. Prepare fresh prior to the experiment.
6. Live-cell imaging.

   Video inverted microscope Nikon Ti-E, Camera: CCD 1392 × 1040.

   CoolSnap HQ2, dry objective: 10× CFI Plan Fluor, NA 0.3.

3 Methods

3.1 Micropatterning

1. Illuminate the coverslips with the deep UV lamp for 5 min to activate them and allow coating with PLL-PEG (*see* **Note 2**).
2. Put 100 μL drops of 0.1 mg/mL PLL-PEG on a parafilm sheet.
3. Place the coverslips on the drops. The side of the coverslip that was illuminated should be in contact with the drop. Incubate for 30 min at RT.
4. Wash the coverslips twice with PBS and once with ddH_2O. Leave the coverslips dry on a tissue wiper, coated side on the top.
5. Wash the mask with 70% ethanol and air-dry it.
6. Put 5 μL drops of ddH_2O on the metal part of the mask.
7. Place the coverslips on the drops. The PLL-PEG coated side of the coverslips should be in contact with the drops.
8. Put the mask with the coverslips in the UV oven with the coverslips facing the opposite way of the lamp. Illuminate for 5 min.
9. Detach the coverslips from the mask using ddH_2O. Dry the coverslips.
10. Micropatterned coverslips are stored at 4 °C, with the PLL-PEG coated side on the top. Avoid aggregation of the coverslips. The coverslips can be stored up to a couple of months.

3.2 Sample Preparation for Live-Cell Microscopy

1. Micropatterned coverslips are placed in a 12-well plate and sterilized using the low intensity UV lamp of the laminar flow hood for 5 min not to destroy the patterns (*see* **Note 3**).
2. Micropatterned coverslips are coated with 1% matrigel diluted in PBS at 37 °C for 15 min.
3. To reach 80–100% confluency, seed 80,000 myoblasts per well.
4. After checking myoblasts are well seeded on patterned lines (*see* **Note 4**), change medium for the differentiation medium for 4 days.
5. Incubate the differentiated myotubes in DMEM medium containing 5 μg/mL calcein-AM (1×) and 50 μg/mL DAPI (1×) for 15 min at 37 °C in the dark. Wash once with PBS and add differentiation medium to wash out the excess of calcein-AM and DAPI.

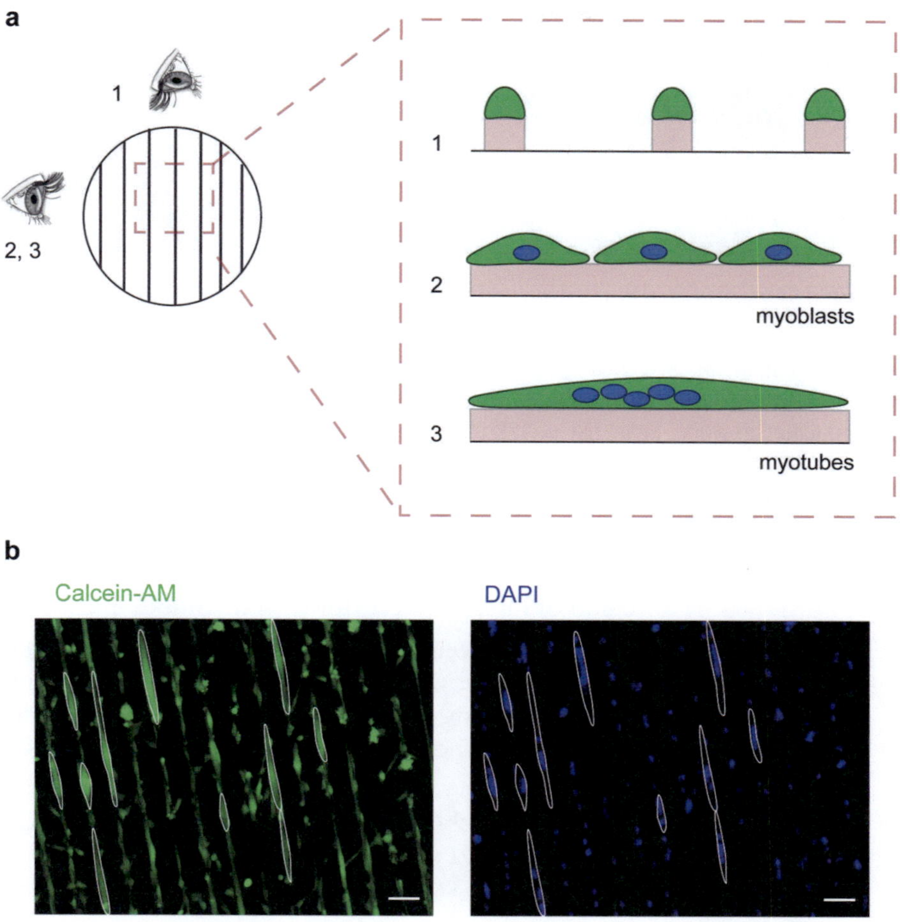

Fig. 1 Human myotubes on line micropatterned coverslips. (**a**) Cartoon of two different side views of myoblasts [1, 2] or myotubes [1, 3] grown on coverslips coated with matrigel on micropatterned line (pink). (**b**) Images of myotubes grown on line micropatterns. Calcein-AM (green) is used to visualize the live cells, and DAPI (blue) allows to select only myotubes, which are outlined in white. Myotubes corresponds to cells having at least two nuclei. Scale bar = 120 μm

Calcein-AM allows the visualization of the living cells and DAPI allows the specific monitoring of myotubes, which correspond to cells containing at least two nuclei (Fig. 1) (*see* **Note 5**).

3.3 Hypo-osmotic Shock and Live-Cell Imaging

1. Place the plate containing the micropatterned myotubes under a videomicroscope (Inverted microscope Nikon Ti-E with CCD camera and 10x air objective) at 37 °C under 5% CO_2.
2. Switch the medium to the hypo-osmotic shock medium (30 mOsm), containing PI (1×) (*see* **Note 6**).
3. Take a picture immediately after changing the medium and every minute for 10 min. DAPI, calcein-AM, and PI emit in the blue, green and red channels respectively (*see* **Note 7**).

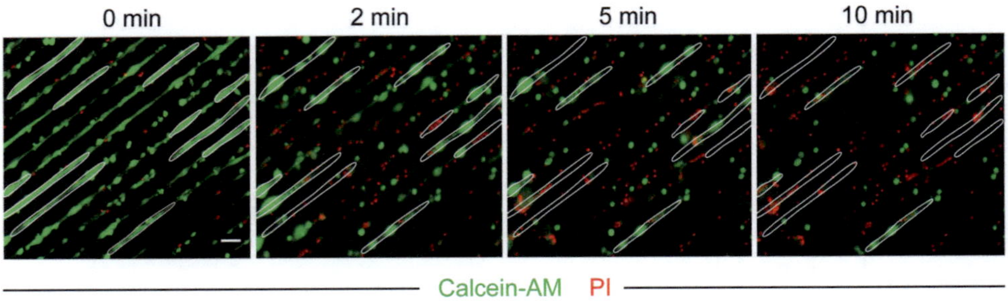

Fig. 2 Membrane integrity in human myotubes. Micropatterned human myotubes were loaded with calcein-AM (green). The medium was switched to hypo-osmotic medium (30 mOsm) supplemented with propidium iodide (PI, red). Representative pictures were taken at the indicated times during hypo-osmotic shock. Myotubes are outlined in white. Scale bar = 120 μm

3.4 Image Analysis

1. Open acquired images with ImageJ software.
2. Make a stack of images for each channel and merge the DAPI and the calcein-AM stacks to identify the myotubes. Define a region of interest (ROI) for each myotubes in the field.
3. Merge the calcein-AM and the PI stack of images. Target myotubes using the ROI defined with the DAPI signal. Each myotubes having a red (PI) signal at the end of the 10 min hypo-osmotic shock is considered as a myotube that has burst (Fig. 2).
4. For each myotube, the time of burst should correspond to the appearance of PI signal in myotube nuclei.

4 Notes

1. Micropatterning is suitable for deformable surfaces such as polydimethylsiloxane (PDMS). But in our hands, it was very difficult to obtain a good differentiation of myotubes and minimal detachment of cells.
2. Note that calcein exists in different colors, which is convenient when used in transfected cells.
3. The use of micropatterns, although making experiment closer to muscle physiology and the analysis easier, can have drawbacks. After myoblast seeding, you first need to make sure that the patterns were done correctly. It can happen that cells are all over the coverslip, which means patterns are not properly done. Patterns can also be difficult to sterilize and contamination often happen during myotube differentiation. Depending on the myoblast cell lines you work with, differentiation into myotubes on micropatterns can be challenging, as less contact zones exist between myoblasts for fusion in these constrained environments. In addition, you need to be careful when changing medium, as myotubes detach easily from their substrate.

4. If at the beginning of the experiment, you have a very hard time finding correctly differentiated myotubes, do not proceed with the experiment, as the remaining myotubes present are generally weaker and have a higher amount of membrane damage than usual in these conditions.
5. Calcein-AM could be used alone, as its disappearance should reflect membrane damage. However, it happens very often that you can observe PI signal in the nuclei before observing the loss of calcein-AM fluorescence, which still indicates that membrane damage has occurred but we would not know it with the simple observation calcein-AM intensity. In addition, it happens sometimes that a rapid membrane repair occurs after damage, in which case we can observe both PI and calcein-AM fluorescence until the end of the time course of the experiment.
6. This technique was used for different types of cells such as muscle cells, endothelial cells, and melanocytes. We also used it to study non-caveolae dependent mechanoprotection, such as in human muscle cells from patients bearing dysferlin mutations [10, 11]. For all these types of cells, only a severe hypo-osmotic shock allowed observing membrane damage. This could be due to the fact that these cells are particularly used to undergo a mechanical stress and are well adapted to these conditions.
7. As the cells undergo a severe hypo-osmotic shock, they swell and tend to move during the experiment. To overcome this issue, you can either use a perfect focus module that you can find on some microscopes. One other solution is to use a plugin, StackReg, available on imageJ, which allows realigning the objects within a stack.

Acknowledgements

The authors would like to thank Christophe Lamaze and all the people from the Membrane dynamics and mechanics of intracellular signaling laboratory. The facilities as well as scientific and technical assistance from staff in the PICT-IBiSA/Nikon Imaging Centre at Institut Curie-CNRS and the France-BioImaging infrastructure (N°ANR-10-INSB-04) are acknowledged. This work was supported by institutional grants from the Curie Institute, INSERM and CNRS, and by specific grants from Association Française contre les Myopathies (AFM): CAV-MUT (17151) to M.D; CAV-STRESS-MUS (14266 to C.M.B), The Lamaze team, the PICT-IBiSA/Nikon Imaging Centre at Institut Curie-CNRS and the France-BioImaging infrastructure are members of Labex CelTisPhyBio (N°ANR-10-LBX-0038) and of IDEX PSL (N°ANR-10-IDEX-0001-02 PSL).

References

1. Lamaze C, Tardif N, Dewulf M et al (2017) The caveolae dress code: structure and signaling. Curr Opin Cell Biol 47:117–125
2. Cheng JPX, Nichols BJ (2016) Caveolae: one function or many? Trends Cell Biol 26:177–189
3. Sinha B, Köster D, Ruez R et al (2011) Cells respond to mechanical stress by rapid disassembly of caveolae. Cell 144:402–413
4. Lo HP, Nixon SJ, Hall TE et al (2015) The caveolin-cavin system plays a conserved and critical role in mechanoprotection of skeletal muscle. J Cell Biol 210:833–849
5. Han B, Copeland CA, Kawano Y et al (2016) Characterization of a caveolin-1 mutation associated with both pulmonary arterial hypertension and congenital generalized lipodystrophy. Traffic 17:1297–1312
6. Lim YW, Lo HP, Ferguson C et al (2017) Caveolae protect notochord cells against catastrophic mechanical failure during development. Curr Biol 27:1968–1981
7. Garcia J, Bagwell J, Njaine B et al (2017) Sheath cell invasion and trans-differentiation repair mechanical damage caused by loss of caveolae in the Zebrafish notochord. Curr Biol 27:1982–1989
8. Cheng JPX, Mendoza-Topaz C, Howard G et al (2015) Caveolae protect endothelial cells from membrane rupture during increased cardiac output. J Cell Biol 211:53–61
9. Dewulf M, Köster D, Sinha B et al (2019) Dystrophy-associated caveolin-3 mutations reveal that caveolae couple IL6/STAT3 signaling with mechanosensing in human muscle cells. Nat Commun 10:1974
10. Barzilai-Tutsch H, Dewulf M, Lamaze C et al (2018) A promotive effect for halofuginone on membrane repair and synaptotagmin-7 levels in muscle cells of dysferlin-null mice. Hum Mol Genet 27:2817–2829
11. Dionnet E, Krahn M, Lévy N et al (2015) Exon 32 skipping of dysferlin rescues membrane repair in patients' cells. J Neuromuscul Dis 2:281–290

Chapter 18

Immunofluorescence-Based Analysis of Caveolin-3 in the Diagnostic Management of Neuromuscular Diseases

Andreas Roos, Denisa Hathazi, and Ulrike Schara

Abstract

Immunohistochemistry- and/or immunofluorescence-based analysis of muscular proteins represents a standard procedure in the diagnostic management of patients suffering from neuromuscular diseases such as "Caveolinopathies" which are caused by mutations in the *CAV3* gene encoding for caveolin-3. Human caveolin-3 is a 151 amino acid sized transmembrane protein localized within caveolae, predominantly expressed in cardiac and skeletal muscle cells and involved in a diversity of cellular functions crucial for muscle cell homeostasis. Loss of caveolin-3 protein abundance is indicative for the presence of pathogenic mutations within the corresponding gene and thus for the diagnosis of "Caveolinopathies." Moreover, description of abnormal immunoreactivity findings for the caveolin-3 protein is increasing in the context of other neuromuscular diseases suggesting that profound knowledge of abnormal caveolin-3-expression and/or distribution findings can be decisive also for the diagnosis of other neurological diseases as well as for a better understanding of the biology of the protein. Here, we summarize the current knowledge about the caveolin-3, report on a protocol for immunofluorescence-based analysis of the protein in the diagnostic workup of neuromuscular patients—also considering problems encountered—and confirm as well as summarize already published abnormal histological findings in muscular pathologies beyond "Caveolinopathies."

Key words Caveolinopathy, Caveolin-3, Protein analyses in muscular diseases, Immunofluorescence in muscular diseases, CAV3-related LGMD, Rippling muscle disease

1 Introduction

1.1 Neuromuscular Diseases

Neuromuscular diseases encompass many conditions altering or impairing the proper functioning of skeletal muscles. Hereby, different inherited or acquired diseases can perturb or even impair proper muscle function and cause muscle wasting and weakness as well as muscle fiber degeneration. These pathological conditions can either be directly representing pathologies of the voluntary muscle or be indirectly representing pathologies of the peripheral nervous system or neuromuscular junctions [1]. In the latter context, diseases of the central nervous system can lead to spasticity or some degree of paralysis, depending on the location and the nature

Cedric M. Blouin (ed.), *Caveolae: Methods and Protocols*, Methods in Molecular Biology, vol. 2169,
https://doi.org/10.1007/978-1-0716-0732-9_18,

of the problem with spinal muscular atrophy as a paradigmatic disease of the lower motor neurons and amyotrophic lateral sclerosis as a mixed upper and lower motor neuron disorder [1]. On a general note, for hereditary neuromuscular diseases so far more than 250 genes have been described [2], and regarding the acquired forms, perturbations of the immune system (autoimmune diseases), exposure to toxic agents, critical illness, and cancer are known triggers for muscle fiber degeneration.

1.2 The Dystrophin-Associated Glycoprotein Complex (DGC) and Related Diseases

The DGC represents a highly organized protein-network localized to the plasma membrane of (muscle) cells which is part of the sarcolemma and facilitates the connection between the cellular cytoskeleton and the extracellular matrix (Fig. 1).

The proteins in this complex were first described by Campbell and colleagues [3], designating the components as dystrophin-associated proteins (DAPs) and dystrophin-associated glycoproteins (DAGs). Hereby, several proteins such as dystrophin, syntrophins, and dystrobrevin localize to the cytoplasmic side. In contrast, other components such as α-dystroglycan and laminin localize to the extracellular part of the complex, while other proteins, including the sarcoglycans, β-dystroglycan, sarcospan, dysferlin, and caveolin-3, are transmembrane components of DGC. Interestingly, mutations in the corresponding genes of several of these components are associated with the clinical manifestation of hereditary neuromuscular diseases (Table 1).

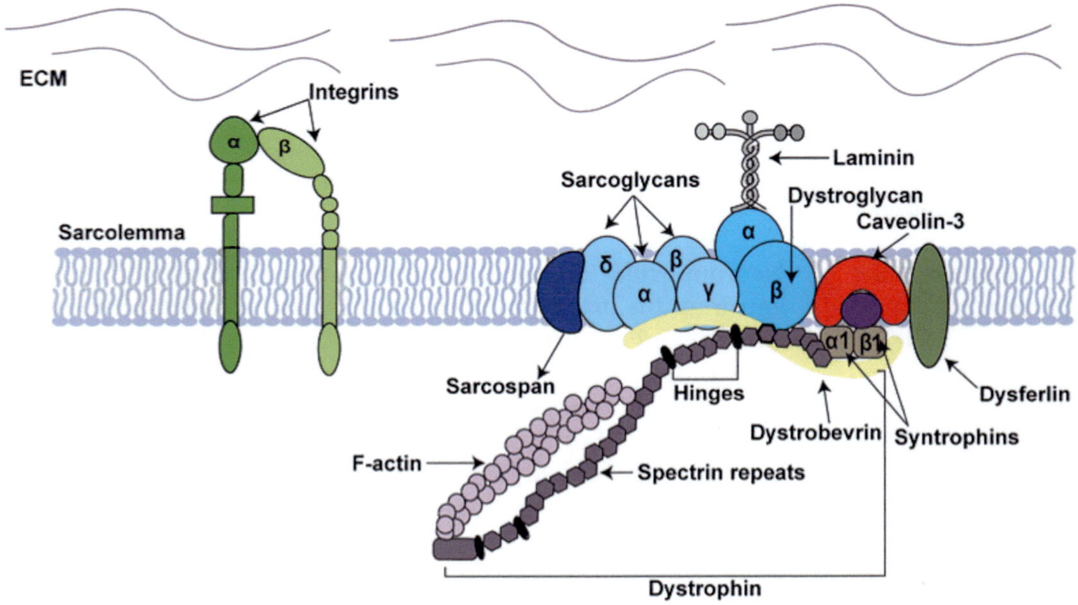

Fig. 1 Schematic depiction of the dystrophin-associated glycoprotein complex (DGC). Caveolin-3 is highlighted in red

Table 1
List of DGC-components associated with neuromuscular diseases (components "exclusively" associated with cardiac diseases have been left out)

Protein	Gene	OMIM # (phenotypes)	Mode of inheritance
Caveolin-3	*CAV3*	192600, 123320, 611818, 614321, 606072	Autosomal dominant and recessive
Dysferlin	*DYSF*	254130, 253601, 606768	Autosomal recessive
Dystroglycans	*DAG1*	616538, 613818	Autosomal recessive
Dystrophin	*DMD*	300376, 302045, 310200	X-chromosomal
Laminin	*LAMA2*	607855, 618138	Autosomal recessive
Alpha-sarcoglycan	*SGCA*	608099	Autosomal recessive
Beta-sarcoglycan	*SGCB*	604286	Autosomal recessive
Gamma-sarcoglycan	*SGCG*	253700	Autosomal recessive
Delta-sarcoglycan	*SGCD*	606685, 601287	Autosomal recessive
Epsilon sarcoglycan	*SGCE*	159900	Autosomal dominant

1.3 Caveolae, Caveolin-3, and Caveolinopathies

Although acting in concert toward a proper connection between cytoskeleton and extracellular matrix in striated muscle fibers, the different DGC components mentioned above also present with individual functional characteristics: caveolins are integral membrane components and represent the major coat proteins of caveolae which are flask-shaped plasma membrane invaginations found in numerous cell types including smooth and striated muscle [4, 5]. Regularly, caveolae present as lipid rafts enriched in clusters of cholesterol, sphingolipids, and glycosyl phosphatidylinositol-anchored proteins [6]. The inverted omega shape of caveolae serves as a membrane reservoir antagonizing mechanical stress burden (by flattening into the membrane to help cells withstand mechanical stress), thus declaring caveolae as mechanoprotective domains [7]. This physiological feature represents an important aspect, especially in cells constantly exposed to mechanical stress such as muscle fibers.

So far, three types of caveolin proteins have been described from which caveolin-3 (encoded by the *CAV3* gene) is predominantly expressed in skeletal and cardiac muscles. However, there is also evidence for caveolin-3 expression in Schwann cells as a patient suffering from a limb-girdle muscular dystrophy complicated by signs of perturbed myelination of peripheral axons has been described [8] and caveolin-3 has also immunohistochemically been detected in Schwann cell caveolae, although with slight

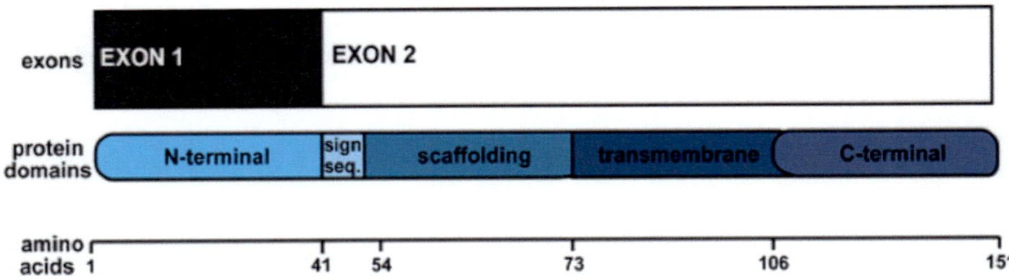

Fig. 2 Structure of *CAV3* gene as well as of the corresponding Caveolin 3 protein divided into functional domains

immunoreactivity [9]. Human caveolin-3 is 151 amino acid sized protein divided into four separate domains: the N-terminal domain, the scaffolding domain, the transmembrane, and the C-terminal domain [10] (Fig. 2). Topological studies revealed that the N-terminal domain undergoes a large-scale pH-mediated rearrangement between soluble and membrane-anchored forms, an important aspect as approximately one-third of pathogenic *CAV3* mutations influence charged residues located within this domain thus suggesting that this transition is likely to be relevant to the molecular etiology of "Caveolinopathies" [11]. The scaffolding segment is responsible for the homo-oligomerization of caveolin-3 which starts in the endoplasmic/sarcoplasmic reticulum where detergent-resistant caveolar complexes are formed. These complexes are afterward transported to the golgi apparatus and exit bearing different lipids. Then, these molecules fuse with the plasma membrane forming a hairpin loop, the so-called caveolae (see above). In mature myofibers, caveolin-3 is present through the T-tubule system where the protein clusters at the necks of the tubule thus facilitating the transmission of the contractile impulses [12]. Caveolin-3 interacts directly with G-protein alpha subunits and can functionally regulate their activities. Moreover, caveolin-3 has shown to regulate voltage-gated potassium channels and plays a role in repair mechanisms of the plasma membrane in both skeletal muscle fibers and cardiomyocytes [13]. Additionally, there is functional evidence that caveolin-3 mediates the caveolae-recruitment of cavin2 and cavin3, proteins involved in caveolar biogenesis and morphology [14]. Further functions in muscle fiber maintenance include regulation of signal transduction, lipid- and glycometabolism, insulin and growth factor signalling, vesicle trafficking, mechanosensitive channel functions, and autophagy [12, 15–19].

Remarkably, under physiological conditions, caveolin-3 was found to be enriched at the neuromuscular junctions (NMJ) where it is associated with the AChR and directly interacts with the MuSK receptor upon agrin stimulation [20]. The exact underlying mechanisms by which caveolin-3 regulates NMJ-integrity are still elusive but it is thought that besides concentrating the AChR receptor at the NMJs due to its scaffolding properties, caveolin-3 is also activating MuSK in a Rac1 depended manner [21]. In addition, it is suggested that the activation of MuSK via self-phosphorylation is decreased in myotubes lacking caveolin-3 [20].

As indicated in Table 1, mutations within the two-exon spanning *CAV3* gene (located at 3p25.1) are leading to muscular diseases named "Caveolinopathies" which are (with exception of only a few recessive mutations) transmitted in an autosomal dominant fashion. So far, more than 40 pathogenic dominant *CAV3* mutations have been described [22]. Pathogenic mutations are mostly amino acid substitutions but also other point mutations such as splice site mutations, leading to the clinical manifestation of muscle disease, have been identified [12]. In addition, there is evidence that some amino acid substitutions within caveolin-3 do not or only slightly change the cellular distribution and functional properties of the protein [23, 24] and may act as modifying factors in the presence of additional stress burden or the presence of further genetic alterations resulting in the clinical manifestation of muscle disease [25, 26].

Clinically, mutations in the *CAV3* gene can result in four distinct, sometimes overlapping phenotypes: limb-girdle muscular dystrophy (LGMD), rippling muscle disease (RMD), distal myopathy (DM), and isolated hyperCKemia (HCK) as well as familial hypertrophic cardiomyopathy [22]. In addition, muscle cramps, myalgia, and myotonia have been linked to "Caveolinopathies" [24, 27–29]. *CAV3* mutations associated with exercise intolerance and rhabdomyolysis further expand the phenotypic spectrum of [24]. Moreover, *CAV3* mutations have been described in the context of genetic double-trouble leading to the manifestation of "overlapping syndromes" [26].

DM is a relatively rare subtype of "Caveolinopathies," and patients experience muscle wasting and weakness, mostly in the small muscles of hands and feet but hypertrophy of the calf muscles can also occur. Muscle biopsy specimen present with a mild variation in fiber size and centralized nuclei [30].

Caveolin-3-related LGMD describes a childhood or adult-onset muscle disease characterized by muscle weakness and wasting affecting mostly the limb and proximal musculature. Muscle biopsy specimens usually show variable fiber size, degenerating/regenerating muscle fibers, centralized myonuclei, and a very mild substitution of muscle with connective tissue. Several mutations affecting the N-terminal (p.N33K, p.V43E), scaffolding (p.T63P, p.G56S, p.C72W, p.W58R), or the membrane-spanning domain (p.A93T, p.P104L, p.T78M, p.T78K) have been linked to a limb-girdle muscular phenotype (formally LGMD1C) [10, 12, 23, 31].

RMD is a muscle disorder with variable age of onset characterized by increased muscle irritability. Muscle biopsy specimen derived from "Caveolinopathy-RMD" patients present with an increase in muscle fiber size variability, mild type I fiber predominance, and internalized nuclei. Patients present with increased fatigability, tiptoe walking, myalgia, and muscle stiffness [32] and hypercholesterolemia has also been described [33]. Leading symptoms possibly starting in childhood are percutan-induced rapid contractions (PIRCs), mounding and rippling [34].

Several *CAV3* mutations associated with RMD, affecting the N-terminal (p.R27Q, p.P28L, p.E46K), scaffolding (p.S52G, p. W70X), or the membrane- spanning domain (p.L85P, p. F103_F104del) have been described [10]. Muscle MRI of caveolin-3-related RMD patients revealed most commonly a vulnerability of rectus femoris and semitendinosus muscles and a further involvement extended to biceps femoris, and gracilis with disease progression or increase in the severity of symptoms. Notably, similar MRI patterns of vulnerability could be observed on reviewing muscle MRI-data of various previously reported phenotypes of "Caveolinopathies" [35].

Some *CAV3* mutations have been identified in patients affected by isolated HCK without clinical symptoms. Usually, they solely present CK-plasma levels four to ten times higher than healthy individuals and histological muscle analysis typically reveals no pathological abnormalities.

Based on the considerable clinical continuum and varying clinical presentations of "Caveolinopathies," one might speculate that additional unknown loci must exist which impacts on the clinical manifestation [12, 30, 36]. This assumption is also supported by the fact that even identical amino acid substitutions can lead to different muscular phenotypes [23, 24, 37].

Mutations in genes encoding for DGC, in most of the cases, are leading to the loss of the affected protein. Given that some of the DGC components physically interact (e.g. dysferlin and caveolin-3), loss of one particular protein might also impinge on the localization or stability of interaction partners as previously described for caveolin-3 abundance in some "Dysferlinopathy" patients [38]; see below. This pathophysiological effect represents an important aspect in the diagnostic management of patients with muscular diseases and has to be considered in the interpretation of immunoblot, immunohistochemistry, and immunoblot-data.

1.4 Immunostaining of Caveolin-3, a Common Procedure in the Diagnostic Management of Myopathic Patients

In spite of dominant inheritance and thus only affecting one allele, these amino acid substitutions very often result in a considerable decrease or even almost complete loss of caveolin-3 within muscle fibers. This pathobiochemical finding represents the consequence of irregular homo-oligomerization of mutant and wild-type caveolin-3. These protein complexes stack in the golgi apparatus and cause stress in the endoplasmic/sarcoplasmic reticulum—golgi continuum associated with the activation of the unfolded protein response [39, 40] (Fig. 3). The irregular mutant-wild-type

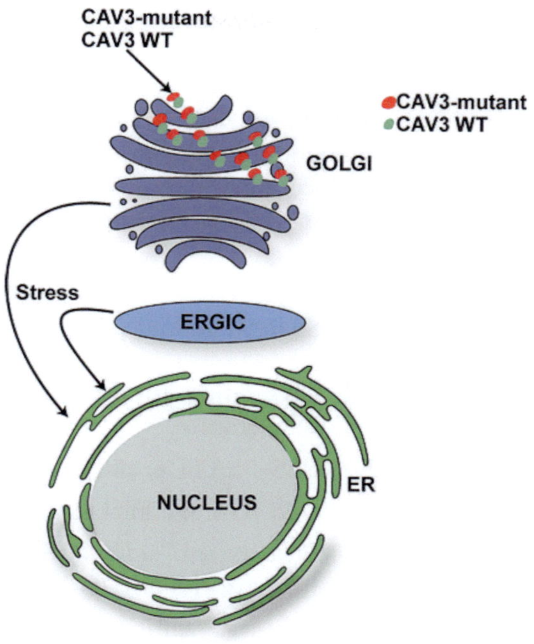

Fig. 3 Schematic depiction of pathophysiological mislocalization of mutant caveolin-3 forming irregular complexes with the wild-type protein (WT) within the golgi causing stress within the ER-golgi intermediate compartment (ERGIC) and the endoplasmic reticulum (ER)

caveolin-3 complexes are transported across the membrane of the golgi apparatus and targeted for degradation by the proteasome leading to the loss of both mutant and wild-type caveolin-3 [41]. This aspect is of significant importance for the immunoblot and/or immunohistochemistry/immunofluorescence-based diagnostic management of patients presumably suffering from "Caveolinopathies" as loss of caveolin-3 or a very considerable decrease (of sarcolemmal localization) indicate the presence of pathogenic CAV3 mutations. Hence, investigation of caveolin-3 protein level represents an important and predictive tool in the diagnostic management of patients with muscular diseases and can be used as a prescreening procedure to narrow-down the clinical subtype (e.g. of LGMDs) or even directly hint toward a CAV3-related muscle disease.

2 Materials

All solutions should be prepared using ultrapure water (prepared by purifying deionized water to attain a sensitivity of 18 MΩ-cm at 25 °C) and analytical grade reagents. Preparation and storage of all reagents should be carried out at room temperature (unless indicated otherwise). Diligently follow all waste disposal regulations when disposing waste materials.

2.1 Muscle Biopsy and Sample Storage

Muscle biopsies are taken either as "open" or as needle biopsies and afterward immediately transferred to the laboratory where glycerine-based embedding of the biomaterial is carried out utilizing Tissue-Tek® O.C.T.™ Compound (Sakura). Notably, this embedding medium completely dissolves during the staining procedure (see below). Prior cryosectioning, samples are stored at −80 °C in a laboratory freezer (*see* **Note 1**).

2.2 Cryostat and Slides

1. Cryostat: thermo Scientific HM525.
2. Slides: Ultraplus, Thermo Scientific.

2.3 Staining

1. PBS; adjust the pH with 1 M NaOH to 7.6: store at room temperature.
2. PBS-BSA (1%) and (3%) solutions.
3. Add BSA at a final concentration of 1% or 3% (w/v) in PBS.
4. Primary antibody: anti-caveolin-3 antibody, BD Transduction Laboratories: #610420.
5. Primary antibody: anti-spectrin antibody, Novocastra NCL-SPEC1.
6. Secondary antibody: goat anti-mouse Alexa 488, Invitrogen: #A-11029.
7. Secondary antibody: goat anti-rabbit Alexa 549, Invitrogen: #A-11032.
8. Fluorescence mounting medium: DAKO fluorescence mounting medium: #S3023.
9. Kimwipes, KIMTECH: #107281.
10. Incubator.

2.4 Microscopy

1. Microscope: Zeiss Axioplan;
2. Camera: Zeiss Axio-Cam ICc1.

3 Methods

3.1 Diagnostic-Based Immunofluorescence-Staining of Caveolin-3 in Patient-Derived Muscle Biopsy Sections

1. Section O.C.T.™ -embedded muscle samples on a cryostat (*see* **Note 1**). Place 5 μm sections of patient biopsies along with sections of muscle biopsies serving as "healthy control" on slides (*see* **Note 2**). Dry the slides at 25 °C for 1 h (incubator).
2. After drying of the section, rehydrate in 1 × PBS for 5 min. Drain excessive PBS with a kimwipe without touching the muscle tissue (*see* **Note 3**).
3. Dilute the primary anti-caveolin-3 antibody 1:500 and the anti-spectrin antibody 1:100 in PBS-BSA 1% (serving as blocking agent) to visualize the plasma membrane (sarcolemma) of

muscle fibers within the section. Add 100 μL of the diluted primary antibody mixture to the muscle section with subsequent incubation for 1 h at 25 °C (incubator) (*see* **Note 4**). Ensure that the utilized primary antibodies are not expired (*see* **Note 6**). Prior samples are incubated with the secondary antibodies, wash slides three times for 30 s in PBS to remove the primary antibodies. Drain excessive PBS with a kimwipe by carefully placing the same one adjacent to the section without touching the muscle tissue (*see* **Note 3**). Dilute secondary antibodies 1:500 in PBS-BSA 1%. Add 100 μL of the diluted secondary antibody mixture to the muscle section with subsequent incubation for 1 h at 25 °C (incubator). Ensure that the utilized secondary antibodies are not expired (*see* **Note 5**). From this step forward, protect samples from light. After the incubation with the secondary antibody solution, wash slides three times for 30 s in PBS.

4. Mount sections with fluorescence mounting medium to visualize myonuclei and protect samples from light until microscopic inspection (*see* **Note 6**). Prior focussing on patient-derived section, proof proper staining and thus successful application of the described protocol by examining the immunolabeling of caveolin-3 and spectrin in the control sections which are placed adjacent to the patient-derived sections on the same slide. Investigate a minimum of 50 muscle fibers (in addition to the H&E staining of the respective muscle biopsy) to draw a conclusion regarding altered abundance and localization of caveolin-3 (*see* **Note 7**).

3.2 Interpretation of Immunostaining Findings: Usual and Unusual Results

Given that the majority of dominant pathogenic mutations result in a degradation of the mutant along with the wild-type protein, overall caveolin-3 level is in the majority of cases reduced in muscle biopsy specimen derived from "Caveolinopathy" patients (*see* description in Subheading 1.4). This biochemical effect represents a benefit in the "immune-based" diagnosis of a "Caveolinopathies" as it makes the generation of caveolin-3 peptide antibodies targeting sequences (commonly) affected by dominant pathogenic amino acid sequence alterations (missense mutations) dispensable and thus also allow the biochemical based identification of "Caveolinopathy" patients harboring so far undescribed mutations leading to novel amino acid substitutions. However, one should keep in mind that not all pathogenic caveolin-3 mutations result in altered abundance or distribution of the protein [23, 24], thus making the immunofluorescence-based diagnosis in such cases challenging.

Here, we exemplarily present the diagnostic-based caveolin-3-immunofluorescence findings in two patients with genetically proven "Caveolinopathies," and seven patients suffering from other molecular genetically confirmed muscle diseases including

ANO5-, *DYSF*-, *SGCA*, *SGCB*-, *SGCG*-related LGMD, *RYR1*-related recessive congenital myopathy, and *DNM2*-related dominant centronuclear myopathy [42]. Moreover, one exemplary case of juvenile dermatomyositis is shown.

Our immunofluorescence-findings confirm the profound reduction of caveolin-3 immunoreactivity (whereas sarcolemma spectrin-immunoreactivity remains normal) in the case of the presence of pathogenic mutations as exemplified in the quadriceps muscle biopsies of both cases presented (Fig. 3). In addition, in *CAV3*-patient 1, a degenerating fiber exhibits sarcoplasmic deposits immunoreactive for caveolin-3 and spectrin and most likely—in accordance with the known pathophysiology—representing protein aggregates targeted for degradation (Fig. 3).

Notably, reduced level of caveolin-3 has been described in some "Dysferlinopathy" patients most likely as a pathophysiological effect of the interrupted functional interaction of both proteins as functional DGC components [43]. Our caveolin-3 immunofluorescence studies of two "Dysferlinopathy" patients did not reveal a reduced immunoreactivity but a diffuse sarcoplasmic distribution in few degenerating fibers as well as frequently the presence of sarcolemmal, subsarcolemmal, and sarcoplasmic dots immunoreactive for caveolin-3 and to a minor degree also for spectrin. Representative findings are shown for one of these two cases in Fig. 3. Given that protein aggregation has already been described in "Dysferlinopathy" patients [44], these deposits most likely also represent protein aggregates formed as part of the pathophysiology upon the presence of pathogenic *DYFS*-mutations. However, further systematic co-staining studies are required to address the question whether these deposits are also immunoreactive for mutant dysferlin. Interestingly, our immunofluorescence studies in the quadriceps muscle biopsy derived from an *ANO5*-patient revealed—similar to the "Dysferlinopathy" patients—a diffuse sarcoplasmic caveolin-3 (but not spectrin) distribution in few degenerating myofibers (Fig. 4). Further caveolin-3 immunofluorescence studies focussing on patients with "Sarcoglycanopathies" revealed no obvious changes in the muscle biopsies of three *SGCA*-patients (one representative case is shown in Fig. 3), but again a diffuse sarcoplasmic distribution in few degenerating fibers along with immunoreactivity of the rim of vacuoles and occasionally sarcolemmal gaps of immunoreactivity for both caveolin-3 and spectrin in three patients with *SGCB*-related LGMD (one representative case is shown in Fig. 3). The latter finding is presumably indicative for the disintegration of the sarcolemma based on disruption of a functional DGC and leading to a leak of the muscle fiber membrane. In three patients with *SGCG*-related LGMD, subsarcolemmal and sarcoplasmic deposits immunoreactive for caveolin-3 (and to a minor degree for also for spectrin) could be observed. In this context, it is important to note that in *VCP*-patient muscle (protein

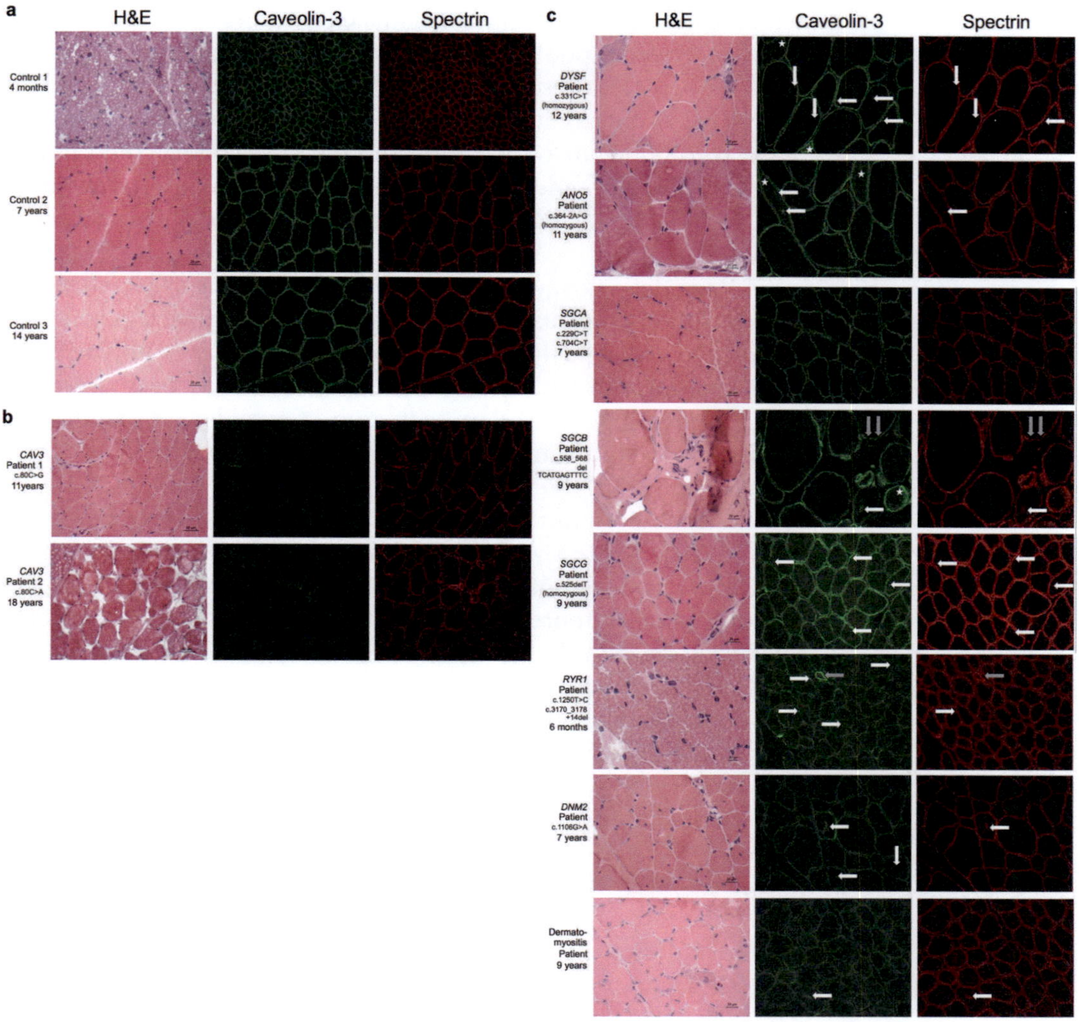

Fig. 4 Caveolin-3 and spectrin immunofluorescence findings in patients with genetically proven muscular diseases including "Caveolinopathies" as well as juvenile dermatomyositis: (**a**) regular caveolin-3 immunoreactivity in muscle biopsy specimen derived from the controls of different ages. (**b**) Absence of caveolin-3 immunoreactivity but not spectrin in two patients of "Caveolinopathy" presenting with LGMD. *CAV3*-patient 1 in addition shows sarcoplasmic deposits immunoreactive for caveolin-3 and spectrin in one degenerating muscle fiber. (**c**) Caveolin-3 and spectrin immunofluorescence studies showed diffuse sarcoplasmic caveolin-3 distribution in a degenerating fiber (white star) and deposits immunoreactive for caveolin-3 and to a minor degree for spectrin (white arrows) in a *DYSF*-patient. Same studies in an *ANO5*-patient also showed diffuse sarcoplasmic caveolin-3 distribution throughout degenerating muscle fibers (white stars) accompanied by caveolin-3 immunoreactivity of vacuoles within one of those degenerating fibers (white arrows). Caveolin-3 and spectrin immunofluorescence studies in a series of "Sarcoglycanopathy" patients revealed no abnormal findings in patients with *SGCA*-mutations (one representative case is shown), whereas again a degenerating fiber presented with diffuse sarcoplasmic caveolin-3 staining (white star), positive caveolin-3 and spectrin immunoreactivity of the rim of a vacuole (white arrows) and lacks of immunoreactivity for both proteins in a diseased fiber (gray arrows). Immunofluorescence analysis of both proteins in a quadriceps muscle biopsy of a representative *SGCG*-patient frequently showed sarcoplasmic dots immunoreactive for caveolin-3 and to a minor degree also for spectrin. Increased sarcolemma-immunoreactivity of both proteins could be observed in

aggregation is here a clinical hallmark), accumulation of caveolin-3 in sarcoplasmic pools along with delocalization from the sarcolemma has been described [45]. Although we could observe the presence of intracellular caveolin-3-positive deposits, a concomitant (profound) delocalization from the sarcolemma could not be observed in the series of our investigated patients suffering from *DYSF*-, *ANO5*- *SGCA*-, *SGCB*-, and *SGCG*-related muscle diseases (Fig. 3).

Regarding further altered caveolin-3 findings in patients with different muscular diseases, one patient with congenital generalized lipodystrophy type 4 and muscular dystrophy due to polymerase I and transcript release factor (*PTRF*) mutations has been described to present with a secondary reduction of caveolin-3 immunoreactivity [35] and a similar finding has been reported in LGMD1D caused by *DNAJB6* mutations [46]. A similar finding could be obtained in a 6-month-old patient suffering from congenital myopathy based on recessive *RYR1*-mutations (Fig. 3). This pathological finding was accompanied by focal increase or total absence of caveolin-3 immunoreactivity (Fig. 3). These findings suggest that reduced caveolin-3 abundance and/or altered distribution might not only occur in muscular diseases based on the loss of functional caveolin-3 interactions but also due to other mechanisms which still need to be discovered by respective cell biological and biochemical studies.

In contrast, upregulation of caveolin-3 protein has been described in some cases of Duchenne muscular dystrophy (DMD). Thus, tight regulation of caveolin-3 appears to be essential in antagonizing the vulnerability of the sarcolemma or for the regeneration of muscle fibers in the disease progression of DMD [38]. The finding of increased caveolin-3 immunoreactivity in few regenerating muscle fibers of the *RYR1*-patient presented by us (Fig. 3) supports this assumption. Increased caveolin-3-immunoreactivity with focal enrichment at regions near neighboring larger muscle fibers has been described in muscle biopsy specimen derived from "TNFα inhibitor myopathy" patients [46].

In the context of abnormal caveolin-3 immunoreactivity findings suggestive for protein aggregation or vacuole-formation and proteolysis as already presented above, in muscle biopsy specimen

Fig. 4 (continued) a regenerating fiber in the quadriceps biopsy of a *RYR1*-patient (gray arrow). Furthermore, focal lacks of immunostaining as well as focal-enhanced immunoreactivity could be detected for caveolin-3 and to a minor degree for spectrin in some fibers within the same biopsy (white arrows). In the biopsy of a patient with a dominant *DNM2*-mutation, positive immunoreactivity was observed at the rim of a vacuole within one fiber as well as at the sites of pathologically centralized myonuclei (white arrows). Positive immunoreactivity for both proteins at the rim of a vacuole could also be detected in the quadriceps biopsy of a representative case suffering from juvenile dermatomyositis. In the same biopsy, protein-staining moreover revealed frequent sarcoplasmic deposits immunoreactive for caveolin-3 and spectrin derived from patients with genetically proven "Desminopathy," membranes of vacuoles showed immunoreactivity for caveolin-3 [47] and in muscle biopsies of patients with *TRPV4* gene mutations (scapuloperoneal neuronopathy) and acid maltase deficiency, caveolin-3 immunoreactivity could as well be observed at the rim of vacuoles [48, 49]. Irregular aggregation of caveolin-3 has moreover been described in LGMD2T caused by *GMPPB* gene mutations [50] as well as an abnormality in internal muscle fiber architecture in *DNM2*-related centronuclear myopathy [51]; Fig. 3, in *MYOT*-related and in *ZASP*-related myofibrillar myopathies (MFM3 & MFM4; [52, 53]), in *ANO5*-related LGMD2L [54]; Fig. 3, in *AGRN*-related congenital myasthenic syndrome [55], in "regional ischemic immune myopathy" (RIIM; [56]) and oculopharyngeal muscular dystrophy [57] as well as in dermatomyositis (complicated by vascular pathology) [58]; Fig. 3. Notably, in our series of three investigated cases of juvenile dermatomyositis, sarcoplasmic caveolin-3-immunoreactive dots were frequently found (Fig. 4) which represent an irregular distribution of caveolin-3 rather than an artifact (*see* **Note 1**). In contrast, a series of patients suffering from sporadic inclusion body myositis has been tested for different components of the DGC, including caveolin-3 and presented with a normal immunohistochemical reaction to the sarcolemma in all tested cases [59]. In the context of a caveolin-3 immunoreactivity at the rim of or within vacuoles, it is worth noting that caveolin-3 has functionally already been linked to modulation of autophagy [19].

In a myopathy characterized by hexagonally cross-linked tubular arrays, exercise intolerance with exercise-induced myalgia and weakness, these ultrastructural characteristic pathomorphological inclusions revealed immunoreactivity selectively for a caveolin-3-antibody whereby germline mutations in the caveolin-3 encoding gene were excluded [60].

Taking into consideration the role of caveolin-3 in the NMJ (see above), four out of seven patients with immune-mediated rippling muscle disease with Myasthenia Gravis (MG), presented with moderately reduced sarcolemmal caveolin-3 (and dysferlin) immunofluorescence staining in a mosaic pattern. Notably, electron microscopy showed that caveolae were almost completely lost at the sarcolemma in the three biopsies examined but not in endothelium [61]. Moreover, in another cohort of patients suffering from MG, caveolin-3 was presenting with altered immunoreactivity at the sarcolemma concomitant with a patchy distribution within a proportion of muscle fibers [62, 63].

3.3 Further Steps in the Diagnostic Management

A considerable reduction or even complete absence of caveolin-3 immunoreactivity is indicative for the presence of a pathogenic mutation in the corresponding gene. This biochemical finding could—in the context of the diagnostic procedure—be further confirmed by additional immunoblot (also showing reduced or absent caveolin-3 protein level) and/or electron microscopic investigations focussing on caveolae architecture. However, these additional studies are not mandatory and a Sanger sequencing-based analysis of the two-exon spanning *CAV3* gene (along with the exon-intron boundaries) should be performed to ascertain a precise molecular genetic diagnosis. In the case that panel or exome sequencing has been performed in the diagnostic workup of the neuromuscular patient, results of the caveolin-3 immunofluorescence studies can serve as a biochemical measure confirming the pathogenicity of identified sequence variants. Taking into consideration that caveolin-3 immunoreactivity can also show abnormal distribution and amount of the protein in other neuromuscular diseases (see above), results of the immunofluorescence studies can thus also serve to evaluate sequence variants in other genes associated with neuromuscular diseases. In this context, it is important to note that pathological caveolin-3 abundances and distribution have only been made in a number of patients suffering from non-*CAV3*-related muscular diseases and more comprehensive studies focussing on the muscular caveolin-3 abundance and distribution in all different neuromuscular entities are needed.

4 Notes

1. Doubtless, long periods between sampling and freezing might cause protein degradation thus also leading to decrease of proteins of interest such as caveolin-3. Although it might be challenging as sampling and freezing are normally performed in different departments, it is strongly recommended to keep the duration as short as possible to warrant the most reliable results. Moreover, the method of muscle biopsy freezing has a significant impact on the quality of results obtained from immunofluorescence (and further) studies. Notoriously, snap-freezing might cause artifacts such as gaps/hole and rims of these gaps might present with false-positive immunofluorescence after staining. In addition, the storage of the muscle biopsy specimen has a significant influence on quality of the staining results as improper long-term storage (such as −20 °C) might also cause freezing artifacts (including ice-crystal formation) and/or foster protein degradation.
2. Concomitant to the immunofluorescence staining, an Hämalaun & Eosin (H&E) staining of a serial section is required to evaluate the quality of the sample diagnostically analysed as exemplified in Fig. 5, and a repetition of the immunofluorescence staining utilizing fresh cryosections is recommended.

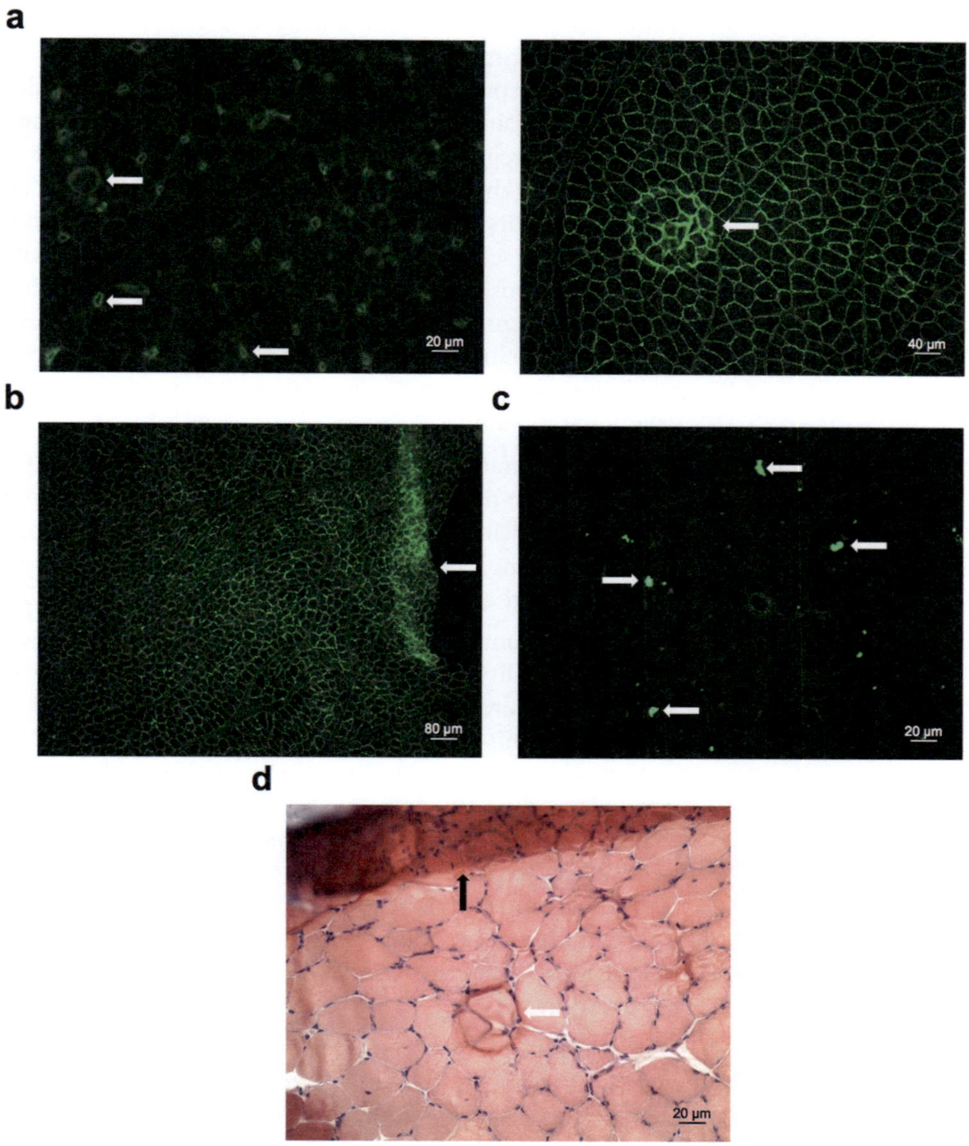

Fig. 5 Illegitimate findings/artifacts in the immunofluorescence-based diagnostic analysis of muscle biopsy sections. (**a**) Deposits of drops in the muscle section showing immunofluorescence labeling (white arrows) and thus influence a proper evaluation of the immunoreactivity of the protein of interest targeted by the primary antibody. (**b**) Irregular folding of the rim of the section of the muscle biopsy making the appearance of increased fluorescence (white arrow). (**c**) Deposits of dirt/dust showing a considerable immunofluorescence-labeling and presumably quenching the immunofluorescence-labeling of the antigens targeted by the primary antibodies. (**d**) Hämalaun & Eosin (H&E) staining of a muscle biopsy section showing a drop (white arrow) as well as irregular folding of the rim of the section (black arrow)

3. It is well known that blurred blades can destroy the integrity of the sections and thus morphology of the muscle tissue consequently making a reliable evaluation of immunuofluorescence staining impossible. Furthermore, deposits of drops in the

muscle biopsy might occur during sectioning. As those drops can show immunofluorescence after staining procedure (Fig. 5), their presence significantly impacts on the reliability of results obtained after immunofluorescence staining. The presence of irregular foldings of the rim of the muscle biopsy (Fig. 5) is a further confounder of immunofluorescence studies and often results as an artifact from improper placement of the muscle biopsy section on the slide. Moreover, deposits of dirt/dust in the muscle biopsy can emerge during sectioning and might significantly complicate data interpretation as they often show immunofluorescence after staining procedure (Fig. 5). A possible source of error is touching the muscle section while drying with a kimwipe.

4. Prolonged durations while staining procedure might cause dry-out of the muscle tissue and thus success of the method in terms of reliability of obtained data. To avoid this undesirable effect, samples must be kept covered in liquid throughout the staining process.
5. Weak or absent immunoreactivity can also be caused by non-functionality of the primary and secondary antibodies. To systematically address this potential error source, proof the shelf life, if this one is not expired but the control muscle sections do not show immunoreactivity, use a higher concentration of antibody or incubate longer. Moreover, double-check that the isotypes of the antibodies used are compatible. In addition, freeze/thaw cycles of the antibodies are detrimental and can cause degradation. It is best to create aliquots of smaller amounts as soon as the product arrives at your location. Inappropriate storage of the antibody might also lead to degradation.
6. One possible error source for presumed weak or absent caveolin3-immunoreactivity might be an incorrect light source/filter set of the microscope. Thus, ensure your microscope is equipped with the correct light source and filter set for the fluorophore chosen. An additional source of error might be that the gain/exposure is too low. To address this potential error source, turn up the gain and/or increase the exposure time to ensure you are capturing any signal present.
7. High background staining often arises as a complication of immunofluorescence studies and can complicate data interpretation. This problem can be caused by autofluorescence. Hence, check to see if there is any fluorescence in an unstained section of the muscle biopsy. Given that inappropriate thickness of the sections might also lead to increased background staining, repetition of the staining protocol by utilizing thinner muscle tissue sections is recommended. Moreover, reduction

of the antibody solutions might reduce the background staining. As secondary antibody sometimes binds nonspecifically, it is crucial to perform an additional staining using the secondary only as control. In the case that unspecific staining can be detected, an alternative secondary antibody should be used for further staining procedures. Insufficient blocking might represent a further error source. In this case, it is recommended to include an additional blocking-step in the proposed protocol or to increase the concentration of the blocking reagent prior to incubation with the primary antibodies. If this is not sufficient, a general change of the blocking reagent should be considered. As proper washing of the muscle biopsy sections between individual steps of the staining protocol is critical, insufficient washing might represent an error source leading to increased background staining. If necessary it is recommended to use a detergent such as NP-40 or TWEEN®.

Acknowledgments

This work was supported by a grant to A.R. from the Deutsche Gesellschaft für Muskelkranke (DGM). We thank Mrs. Swantje Hertel for expert technical assistance.

References

1. Morrison BM (2016) Neuromuscular Diseases. Semin Neurol 36:409–418
2. https://neuromuscular.wustl.edu/. Accessed 25 Feb 2019
3. Ervasti JM, Ohlendieck K, Kahl SD, Gaver MG, Campbell KP (1990) Deficiency of a glycoprotein component of the dystrophin complex in dystrophic muscle. Nature 345:315–319
4. Minetti C, Bado M, Broda P, Sotgia F, Bruno C, Galbiati F, Volonte D, Lucania G, Pavan A, Bonilla E, Lisanti MP, Cordone G (2002) Impairment of caveolae formation and T-system disorganization in human muscular dystrophy with caveolin-3 deficiency. Am J Pathol 160:265–270
5. Razani B, Lisanti MP (2001) Caveolins and caveolae: molecular and functional relationships. Exp Cell Res 271:36–44
6. Williams TM, Lisanti MP (2004) The Caveolin genes: from cell biology to medicine. Ann Med 36:584–595
7. Parton RG (2018) Caveolae: structure, function, and relationship to disease. Annu Rev Cell Dev Biol 34:111–136
8. Brauers E, Dreier A, Roos A, Wormland B, Weis J, Kruttgen A (2010) Differential effects of myopathy-associated caveolin-3 mutants on growth factor signaling. Am J Pathol 177:261–270
9. Kawahara T (2004) Caveolae localization and caveolin expressions in Schwann cells of mature rat spinal nerves. Kurume Med J 51:263–271
10. McNally EM, de Sa ME, Duggan DJ, Bonnemann CG, Lisanti MP, Lidov HG, Vainzof M, Passos-Bueno MR, Hoffman EP, Zatz M, Kunkel LM (1998) Caveolin-3 in muscular dystrophy. Hum Mol Genet 7:871–877
11. Kim JH, Schlebach JP, Lu Z, Peng D, Reasoner KC, Sanders CR (2016) A pH-mediated topological switch within the N-terminal domain of human Caveolin-3. Biophys J 110:2475–2485
12. Gazzerro E, Sotgia F, Bruno C, Lisanti MP, Minetti C (2010) Caveolinopathies: from the biology of caveolin-3 to human diseases. Eur J Hum Genet 18:137–145
13. Cai C, Weisleder N, Ko JK, Komazaki S, Sunada Y, Nishi M, Takeshima H, Ma J (2009) Membrane repair defects in muscular dystrophy are linked to altered interaction

between MG53, caveolin-3, and dysferlin. J Biol Chem 284:15894–15902

14. McMahon KA, Zajicek H, Li WP, Peyton MJ, Minna JD, Hernandez VJ, Luby-Phelps K, Anderson RG (2009) SRBC/cavin-3 is a caveolin adapter protein that regulates caveolae function. EMBO J 28:1001–1015
15. Huang H, Bae C, Sachs F, Suchyna TM (2013) Caveolae regulation of mechanosensitive channel function in myotubes. PLoS One 8:e72894
16. Shang L, Chen T, Deng Y, Huang Y, Huang Y, Xian J, Lu W, Yang L, Huang Q (2017) Caveolin-3 promotes glycometabolism, growth and proliferation in muscle cells. PLoS One 12:e0189004
17. Deng YF, Huang YY, Lu WS, Huang YH, Xian J, Wei HQ, Huang Q (2017) The Caveolin-3 P104L mutation of LGMD-1C leads to disordered glucose metabolism in muscle cells. Biochem Biophys Res Commun 486:218–223
18. Ohsawa Y, Okada T, Kuga A, Hayashi S, Murakami T, Tsuchida K, Noji S, Sunada Y (2008) Caveolin-3 regulates myostatin signaling. Mini-review. Acta Myol 27:19–24
19. Kassan A, Pham U, Nguyen Q, Reichelt ME, Cho E, Patel PM, Roth DM, Head BP, Patel HH (2016) Caveolin-3 plays a critical role in autophagy after ischemia-reperfusion. Am J Physiol Cell Physiol 311:C854–C865
20. Hezel M, de Groat WC, Galbiati F (2010) Caveolin-3 promotes nicotinic acetylcholine receptor clustering and regulates neuromuscular junction activity. Mol Biol Cell 21:302–310
21. Glass DJ, Bowen DC, Stitt TN, Radziejewski C, Bruno J, Ryan TE, Gies DR, Shah S, Mattsson K, Burden SJ, DiStefano PS, Valenzuela DM, DeChiara TM, Yancopoulos GD (1996) Agrin acts via a MuSK receptor complex. Cell 85:513–523
22. Gazzerro E, Bonetto A, Minetti C (2011) Caveolinopathies: translational implications of caveolin-3 in skeletal and cardiac muscle disorders. Handb Clin Neurol 101:135–142
23. Traverso M, Gazzerro E, Assereto S, Sotgia F, Biancheri R, Stringara S, Giberti L, Pedemonte M, Wang X, Scapolan S, Pasquini E, Donati MA, Zara F, Lisanti MP, Bruno C, Minetti C (2008) Caveolin-3 T78M and T78K missense mutations lead to different phenotypes in vivo and in vitro. Lab Investig 88:275–283
24. Scalco RS, Gardiner AR, Pitceathly RD, Hilton-Jones D, Schapira AH, Turner C, Parton M, Desikan M, Barresi R, Marsh J, Manzur AY, Childs AM, Feng L, Murphy E, Lamont PJ, Ravenscroft G, Wallefeld W, Davis MR, Laing NG, Holton JL, Fialho D, Bushby K, Hanna MG, Phadke R, Jungbluth H, Houlden H, Quinlivan R (2016) CAV3 mutations causing exercise intolerance, myalgia and rhabdomyolysis: expanding the phenotypic spectrum of caveolinopathies. Neuromuscul Disord 26:504–510
25. Brauers E, Roos A, Kollipara L, Zahedi RP, Beckmann A, Mohanadas N, Bauer H, Hausler M, Thoma S, Kress W, Senderek J, Weis J (2017) The caveolin-3 G56S sequence variant of unknown significance: muscle biopsy findings and functional cell biological analysis. Proteomics Clin Appl 11
26. Ricci G, Scionti I, Ali G, Volpi L, Zampa V, Fanin M, Angelini C, Politano L, Tupler R, Siciliano G (2012) Rippling muscle disease and facioscapulohumeral dystrophy-like phenotype in a patient carrying a heterozygous CAV3 T78M mutation and a D4Z4 partial deletion: further evidence for "double trouble" overlapping syndromes. Neuromuscul Disord 22:534–540
27. Herrmann R, Straub V, Blank M, Kutzick C, Franke N, Jacob EN, Lenard HG, Kroger S, Voit T (2000) Dissociation of the dystroglycan complex in caveolin-3-deficient limb girdle muscular dystrophy. Hum Mol Genet 9:2335–2340
28. Macias A, Gambin T, Szafranski P, Jhangiani SN, Kolasa A, Obersztyn E, Lupski JR, Stankiewicz P, Kaminska A (2016) CAV3 mutation in a patient with transient hyperCKemia and myalgia. Neurol Neurochir Pol 50:468–473
29. Milone M, McEvoy KM, Sorenson EJ, Daube JR (2012) Myotonia associated with caveolin-3 mutation. Muscle Nerve 45:897–900
30. Chen J, Zeng W, Han C, Wu J, Zhang H, Tong X (2016) Mutation in the caveolin-3 gene causes asymmetrical distal myopathy. Neuropathology 36:485–489
31. Gonzalez-Perez P, Gallano P, Gonzalez-Quereda L, Rivas-Infante E, Teijeira S, Navarro C, Bautista-Lorite J (2009) Phenotypic variability in a Spanish family with a caveolin-3 mutation. J Neurol Sci 276:95–98
32. Van den Bergh PY, Gerard JM, Elosegi JA, Manto MU, Kubisch C, Schoser BG (2004) Novel missense mutation in the caveolin-3 gene in a Belgian family with rippling muscle disease. J Neurol Neurosurg Psychiatry 75:1349–1351
33. Ibarretxe D, Pelleja J, Ortiz N, Masana L (2017) Caveolin 3 deficiency myopathy associated with dyslipidemia: treatment challenges

and possible pathophysiological association. J Clin Lipidol 11:1280–1283

34. Schara U, Vorgerd M, Popovic N, Schoser BG, Ricker K, Mortier W (2002) Rippling muscle disease in childhood. J Child Neurol 17:483–490
35. Ishiguro K, Nakayama T, Yoshioka M, Murakami T, Kajino S, Shichiji M, Sato T, Hino-Fukuyo N, Kuru S, Osawa M, Nagata S, Okubo M, Murakami N, Hayashi YK, Nishino I, Ishigaki K (2018) Characteristic findings of skeletal muscle MRI in caveolinopathies. Neuromuscul Disord 28:857–862
36. Arias Gomez M, Alberte-Woodwar M, Arias-Rivas S, Dapena D, Pintos E, Navarro C (2011) Unilateral calf atrophy secondary to a de novo mutation of the caveolin-3 gene. Muscle Nerve 44:126–128
37. Spadafora P, Liguori M, Andreoli V, Quattrone A, Gambardella A (2012) CAV3 T78M mutation as polymorphic variant in South Italy. Neuromuscul Disord 22:669–670; author reply 670-661
38. Galbiati F, Razani B, Lisanti MP (2001) Caveolae and caveolin-3 in muscular dystrophy. Trends Mol Med 7:435–441
39. Kuga A, Ohsawa Y, Okada T, Kanda F, Kanagawa M, Toda T, Sunada Y (2011) Endoplasmic reticulum stress response in P104L mutant caveolin-3 transgenic mice. Hum Mol Genet 20:2975–2983
40. Gonzalez Coraspe JA, Weis J, Anderson ME, Munchberg U, Lorenz K, Buchkremer S, Carr S, Zahedi RP, Brauers E, Michels H, Sunada Y, Lochmuller H, Campbell KP, Freier E, Hathazi D, Roos A (2018) Biochemical and pathological changes result from mutated Caveolin-3 in muscle. Skelet Muscle 8:28
41. Galbiati F, Volonte D, Minetti C, Bregman DB, Lisanti MP (2000) Limb-girdle muscular dystrophy (LGMD-1C) mutants of caveolin-3 undergo ubiquitination and proteasomal degradation. Treatment with proteasomal inhibitors blocks the dominant negative effect of LGMD-1C mutanta and rescues wild-type caveolin-3. J Biol Chem 275:37702–37711
42. Bonne G, Rivier F, Hamroun D (2018) The 2019 version of the gene table of neuromuscular disorders (nuclear genome). Neuromuscul Disord 28:1031–1063
43. Walter MC, Braun C, Vorgerd M, Poppe M, Thirion C, Schmidt C, Schreiber H, Knirsch UI, Brummer D, Muller-Felber W, Pongratz D, Muller-Hocker J, Huebner A, Lochmuller H (2003) Variable reduction of caveolin-3 in patients with LGMD2B/MM. J Neurol 250:1431–1438
44. Nilsson MI, Laureano ML, Saeed M, Tarnopolsky MA (2013) Dysferlin aggregation in limb-girdle muscular dystrophy type 2B/Miyoshi myopathy necessitates mutational screen for diagnosis (corrected). Muscle Nerve 47:740–747
45. Ritz D, Vuk M, Kirchner P, Bug M, Schutz S, Hayer A, Bremer S, Lusk C, Baloh RH, Lee H, Glatter T, Gstaiger M, Aebersold R, Weihl CC, Meyer H (2011) Endolysosomal sorting of ubiquitylated caveolin-1 is regulated by VCP and UBXD1 and impaired by VCP disease mutations. Nat Cell Biol 13:1116–1123
46. https://neuromuscular.wustl.edu/pathol/dnajb6.htm. Accessed 25 Feb 2019
47. https://neuromuscular.wustl.edu/pathol/tnfaimpp.htm. Accessed 25 Feb 2019
48. https://neuromuscular.wustl.edu/pathol/desmin.htm. Accessed 25 Feb 2019
49. https://neuromuscular.wustl.edu/pathol/spsmatrpv4.htm. Accessed 25 Feb 2019
50. https://neuromuscular.wustl.edu/pathol/acidmalt.htm. Accessed 25 Feb 2019
51. https://neuromuscular.wustl.edu/pathol/gmppb.htm. Accessed 25 Feb 2019
52. https://neuromuscular.wustl.edu/pathol/centnucl.htm. Accessed 25 Feb 2019
53. https://neuromuscular.wustl.edu/pathol/lgd1a.html. Accessed 25 Feb 2019
54. https://neuromuscular.wustl.edu/pathol/zasp.htm. Accessed 25 Feb 2019
55. https://neuromuscular.wustl.edu/pathol/ano5.htm. Accessed 25 Feb 2019
56. https://neuromuscular.wustl.edu/pathol/agrin.htm. Accessed 25 Feb 2019
57. https://neuromuscular.wustl.edu/pathol/par neonec.htm. Accessed 25 Feb 2019
58. https://neuromuscular.wustl.edu/pathol/opmd.htm. Accessed 25 Feb 2019
59. de Camargo LV, de Carvalho MS, Shinjo SK, de Oliveira ASB, Zanoteli E (2018) Clinical, histological, and immunohistochemical findings in inclusion body myositis. Biomed Res Int 2018:5069042
60. Claeys KG, Pellissier JF, Garcia-Bragado F, Weis J, Urtizberea A, Poza JJ, Cobo AM, Stoltenburg G, Figarella-Branger D, Willems PJ, Depuydt CE, Kleiner W, Pouget J, Piraud M, Brochier G, Romero NB, Fardeau M, Goebel HH, Bonnemann CG, Voit T, Eymard B, Laforet P (2010) Myopathy with hexagonally cross-linked crystalloid

inclusions: delineation of a clinico-pathological entity. Neuromuscul Disord 20:701–708

61. Schoser B, Jacob S, Hilton-Jones D, Muller-Felber W, Kubisch C, Claus D, Goebel HH, Vita G, Vincent A, Toscano A, Van den Bergh P (2009) Immune-mediated rippling muscle disease with myasthenia gravis: a report of seven patients with long-term follow-up in two. Neuromuscul Disord 19:223–228

62. Iwasa K, Furukawa Y, Yoshikawa H, Yamada M (2016) Caveolin-3 is aberrantly expressed in skeletal muscle cells in myasthenia gravis. J Neuroimmunol 301:30–34

63. https://neuromuscular.wustl.edu/pathol/mgpath.htm. Accessed 25 Feb 2019

Index

Cedric M. Blouin (ed.), *Caveolae: Methods and Protocols*, Methods in Molecular Biology, vol. 2169,
https://doi.org/10.1007/978-1-0716-0732-9,

MIX
Papier aus verantwortungsvollen Quellen
Paper from responsible sources
FSC® C105338

If you have any concerns about our products,
you can contact us on
ProductSafety@springernature.com

In case Publisher is established outside the EU,
the EU authorized representative is:
Springer Nature Customer Service Center GmbH
Europaplatz 3, 69115 Heidelberg, Germany

Printed by Libri Plureos GmbH
in Hamburg, Germany